New Directions for
Adult and Continuing
Education

Susan Imel
Jovita M. Ross-Gordon
COEDITORS-IN-CHIEF

Authenticity in Teaching

Patricia Cranton

EDITOR

Number 111 • Fall 2006
Jossey-Bass
San Francisco

1-07

#71831775

Authenticity in Teaching
Patricia Cranton (ed.)
New Directions for Adult and Continuing Education, no. 111
Susan Imel, Jovita M. Ross-Gordon, Coeditors-in-Chief

Microfilm copies of issues and articles are available in 16mm and 35mm, as well as microfiche in 105mm, through University Microfilms Inc., 300 North Zeeb Road, Ann Arbor, Michigan 48106-1346.

New Directions for Adult and Continuing Education (ISSN 1052-2891, electronic ISSN 1536-0717) is part of The Jossey-Bass Higher and Adult Education Series and is published quarterly by Wiley Subscription Services, Inc., A Wiley Company, at Jossey-Bass, 989 Market Street, San Francisco, California 94103-1741. Periodicals Postage Paid at San Francisco, California, and at additional mailing offices. POSTMASTER: Send address changes to New Directions for Adult and Continuing Education, Jossey-Bass, 989 Market Street, San Francisco, California 94103-1741.

Subscriptions cost $80.00 for individuals and $195.00 for institutions, agencies, and libraries.

Editorial correspondence should be sent to the Coeditors-in-Chief, Susan Imel, ERIC/ACVE, 1900 Kenny Road, Columbus, Ohio 43210-1090, e-mail: imel.1@osu.edu; or Jovita M. Ross-Gordon, Southwest Texas State University, EAPS Dept., 601 University Drive, San Marcos, TX 78666.

Cover photograph by Jack Hollingsworth@Photodisc

www.josseybass.com

CONTENTS

EDITOR'S NOTES

Authenticity is one of those concepts—like soul, spirit, or imagination—that are easier to define in terms of what they are not than what they are. We can fairly easily say that someone who lies to students or who pretends to know things he or she does not know or who deliberately dons a teaching persona is not authentic. But do the opposite behaviors guarantee authentic teaching? Not necessarily. Becoming an authentic teacher appears to be a developmental process that relies on experience, maturity, self-exploration, and reflection (Cranton and Carusetta, 2004a). It is our purpose in this volume to explore a variety of ways of thinking about authenticity in teaching from the perspective of scholars who dedicate themselves to understanding adult education theory and research and from that of practitioners who see themselves as working toward authentic practice.

The impetus for the volume comes from research a colleague and I are currently engaged in and have been conducting over the past five years (Cranton and Carusetta, 2004b). From interviews with and observation of twenty-two educators in a variety of disciplines, we propose that authenticity consists of five dimensions: self-awareness, awareness of others (especially students), relationships with students, awareness of the educational context and its influence on practice, and critical self-reflection on practice as a way to distinguish oneself from the collective of educators. Each dimension develops over time, as we witnessed by following educators over several years and by contrasting new and experienced educators. Two of the contributors to this volume, Katherine A. Frego (Chapter Four) and Russell Hunt (Chapter Five), are experienced and award-winning educators who were participants in our first three-year research project. Hunt has joined us as a co-investigator as we continue the research. I am pleased to see their perspectives in this volume.

It was my intent to organize this volume around the five facets of authenticity revealed in our research and to explore them from different perspectives. The organization did not turn out to be quite as clear-cut as this statement implies, but neither are the five facets clear-cut or independent of each other, so it seems the final product reflects the inherent messiness of the idea of authenticity.

In Chapter Three, John M. Dirkx addresses the first facet with an in-depth exploration of self-awareness from a Jungian perspective, through the lens of soul work, imagination, and work with the unconscious. In spite of the complexity of the processes he delves into, Dirkx is able to give us practical advice about how to better understand ourselves as human beings and

NEW DIRECTIONS FOR ADULT AND CONTINUING EDUCATION, no. 111, Fall 2006 © 2006 Wiley Periodicals, Inc.
Published online in Wiley InterScience (www.interscience.wiley.com) • DOI: 10.1002/ace.222

hence as teachers. Each of the contributors in some way touches on self-awareness; Lloyd Kornelsen, for example, writes about openness, and Stephen Brookfield about personhood.

Awareness of students—their needs, their humanity, their development—forms the foundation of Frego's Chapter Four, on relationships with students, and it is also a part of Lin's understanding of the cultural dimensions of authenticity (Chapter Six). Frego's primary purpose in her chapter is to comprehensively discuss the nature of caring for and relating to students in order to help them learn and develop. In doing this, she elaborates on the third dimension from our research, the role of relationships with students in being an authentic teacher. Kornelsen, in Chapter Seven, also addresses relationships that are derived from teacher presence in the classroom and the special "flow" that occurs when everything is happening as it should.

Being aware of one's context and its influence on authenticity in teaching is a facet of our research that deserves further exploration. Context includes everything from class size and institutional policies and procedures through issues of class, gender, culture, and race. It is context that leads us to construct our idea of who a teacher is and what he or she does. It is our context that forms values and also prevents us from acting on them. In our interviews with educators, power was often raised as a constraint to authenticity, and as an inevitable part of the context of teaching. Educators have power by virtue of their position and their responsibilities; as Hunt points out, students have power through course evaluations; and as Brookfield and English remind us, power is exercised throughout the teaching and learning context by all participants. In this volume, the context of teaching and the part it plays in developing authenticity is a part of several chapters. In Chapter One, Brookfield addresses the power issue head on, using it as a lens to critique the notion of authenticity. He argues that critical teaching and authentic teaching do not sit well together. In Chapter Two, English considers context in relation to feminist approaches to teaching and learning, particularly how our ideas of women's learning have served to further marginalize women as learners and thereby interfere with authentic teaching and learning. In Chapter Six, Hunt takes a more pragmatic view of context in terms of institutional constraints and helps us see what those constraints are and how we might find alternative ways of working with and around them. Finally, in Chapter Six, Lin examines how cultural contexts lead to our construction of the concept of authenticity. Using vivid examples and quotes from interviews with Asian students in the United States, Lin leads us to see how the values that form authentic behavior are context-bound.

Critical self-reflection, the fifth facet of authenticity in our research (Cranton and Carusetta, 2004b), involves reflection on the others—(self, other, relationships, and context). As such, it is a theme that threads through all the chapters of the volume. Brookfield reflects on congruence, full disclosure, responsiveness, and personhood (indicators of authenticity) through the lens of power; English reflects on authenticity through feminist perspectives. Dirkx, though his writing is based on imagination and soul work, brings in a strong reflective component as he asks us to engage in self-exploration through journal writing and interpretation of images and emotions. In describing her relationships with students, Frego continuously engages in self-reflection on her practice in her writing. Hunt helps us critically examine the institutional context in being authentic; Lin challenges us to see authenticity through the eyes of someone from another culture; and Kornelsen models self-reflection as he works to understand what teaching with presence means.

References

Cranton, P., and Carusetta, E. "Developing Authenticity in Teaching as Transformative Learning." *Journal of Transformative Education,* 2004a, 2(4), 276–293.
Cranton, P., and Carusetta, E. "Perspectives on Authenticity in Teaching." *Adult Education Quarterly,* 2004b, 55(1), 5–22.

Patricia Cranton
Editor

PATRICIA CRANTON *is a visiting professor of adult education at Penn State University at Harrisburg.*

1

Teaching behaviors prized by students as evidence of responsiveness—particularly making full disclosure of our agendas and having our words and actions be consistent—often bring us up against the contradictions and realities of power.

Authenticity and Power

Stephen D. Brookfield

An essential component of skillful teaching—teaching that is grounded in, and responsive to, awareness of how students are experiencing learning—is the attempt to find out how students experience learning and perceive teaching. From the student's perspective, viewing the teacher as both an ally and an authority is an important component of successful learning. Students want to know their teachers stand for something and have something useful and important to offer, but they also want to be able to trust and rely on them. This combination of the teacher as ally and authority is rarely evident in professional protocols or broad standardized indicators of effectiveness. When describing teachers who have made a difference in their lives, or who are recalled as memorable and significant, students rarely use the language of effectiveness. Instead they say they trust a particular teacher to be straight with them, or that a teacher really helped them get something important.

If the teacher is effective, it is because she combines the element of having something important to say, demonstrate, and teach with being open and honest with students. The former quality is that of credibility, the latter the concept of authenticity. Authentic teachers do not go behind a student's back, keep an agenda private, or double-cross students by dropping a new evaluative criterion or assignment into a course halfway through the semester. An authentic teacher is one students trust to be honest and helpful. She is seen as a flesh-and-blood human being with passions, enthusiasms, frailties, and emotions, not as someone who hides behind a collection of learned role behaviors appropriate to the title of professor. From a student's viewpoint, credibility and authenticity need to be recognized in a teacher if the

NEW DIRECTIONS FOR ADULT AND CONTINUING EDUCATION, no. 111, Fall 2006 © 2006 Wiley Periodicals, Inc.
Published online in Wiley InterScience (www.interscience.wiley.com) • DOI: 10.1002/ace.223

5

person is to be seen as an important enhancer of learning—as an authoritative ally, in other words.

Interestingly, it appears that an optimal learning environment is one where both these characteristics are kept in a state of congenial tension. A classroom where teacher credibility is clearly present but authenticity somewhat absent is one where students usually feel their time has been reasonably well spent (because necessary skills or knowledge have been learned) but also one that has been experienced as threatening, intimidating, or unwelcoming. Without authenticity, the teacher is seen as potentially a loose cannon, liable to make major changes of direction without prior warning. On the other hand, a teacher who is seen as authentic but not credible may leave students feeling they have not learned much of importance when it comes to content. Such teachers are spoken of in friendly terms but not viewed as offering much of genuine substance. If learners know they need to develop specific skills or understand certain concepts in order to pass an exam, gain licensure, begin a new career, get out of unemployment, and so on, a teacher with authenticity but no credibility creates great frustration. This is an uncomfortable contradiction for learners to experience.

In this chapter, I address the most common indicators of authenticity that learners identify and then critique them through the lens of power. As we shall see, what learners claim as authentic behavior is sometimes contradicted by the need to exercise power in a way that enhances learning and nurtures change.

Common Indicators of Authenticity

Students recognize that teachers are authentic when the teacher is perceived to be an ally in learning who is trustworthy, open, and honest in dealing with students. Such teachers are viewed as allies in learning because they clearly have the students' interests at heart and wish to see students succeed. In Grimmet and Neufeld's words (1994), authentic teachers strive to do "what is good and important for learners in any given context and set of circumstances" (p. 4) and are perceived this way by learners. This is echoed by the teachers interviewed by Cranton and Carusetta (2004), who speak about the importance of being helpful to learners more than any other factor. However, students see authenticity as more than just being helpful. It also means being viewed as trustworthy. Colloquially, students often say that such teachers "walk the talk," practice what they preach, and have no hidden agenda. Cranton (2001) views this dimension of authenticity as "the expression of one's genuine Self in the community and society" (p. vii). In Palmer's terms (1997), we teach who we are. It is interesting that none of these formulations necessarily implies that students personally like such teachers (though they often do). The most important thing is that the teachers can be trusted. How is such trust developed? In thousands of Critical Incident Questionnaires (CIQs) completed by adult students in my courses and workshops over the past fifteen years, four specific indicators of authenticity are typically mentioned: congruence, full disclosure, responsiveness, and personhood.

Congruence. Congruence here is between words and actions, between what you say you will do and what you actually do. Nothing destroys students' trust in a teacher more quickly than seeing the teacher espouse a set of principles or commitments (for example, to democracy, active, participatory learning, critical thinking, or responsiveness to students' concerns) and then behave in a way that contradicts them. Students usually come to know pretty quickly when they are being manipulated. You may be able to get away with breaking a promise to them once, but that's pretty much it.

Spuriously democratic teachers tell students that the curriculum, methods, and evaluative criteria are open to genuine negotiation and in large measure are in the students' hands. As the course proceeds, however, it becomes clear that the democratically negotiated curricula to be studied, methods to be used, and evaluative criteria to be applied just happen to match the teacher's own preferences. *Falsely participatory teachers* tell students they don't want to lecture too much, value students' contributions, and will use a mixture of teaching approaches (role plays, case studies, simulations, small group discussions, peer-learning triads) that require active student participation. They then proceed to lecture most of the time (each week protesting that this is a temporary necessity because the class is falling behind), not allowing time for questions or not really answering those questions that are raised.

Teachers who are *counterfeit critical thinkers* say they welcome a questioning of all viewpoints and assertions but then bristle when this encouragement is applied to their own ideas and make it clear that certain viewpoints (often those the teacher dislikes) are out of bounds. Practicing *phony responsiveness* happens when teachers profess to respect students' views and then refuse to negotiate around any concerns students raise. In these instances, students quickly conclude that your word is worthless, that any promise you make cannot be taken seriously, and that you are not to be trusted. They may still think they can learn something from you, but they will not experience this as happening in a congenial environment.

The problem is that sometimes we do not realize how incongruent our words and actions appear to students. We may genuinely believe we are living out commitments we made earlier in the course and in the absence of student criticism be completely unaware of how much we are shooting ourselves in the foot. Few students will have the nerve to call you on your lack of authenticity. Mostly they will decide it's simpler not to risk offending you and safer to keep their head down and not make a fuss. So we may be entirely unaware of the impression we're creating.

How can teachers avoid unwittingly falling afoul of the "do as I say, not as I do" trap? Two responses suggest themselves. The first is to use some kind of classroom assessment method to check for perceived inconsistency in your words and actions. My experience is that it is mentioned widely on such instruments as soon as it is perceived to occur. I have sometimes made an off-the-cuff statement that for me was an expression of mild preference but that was taken by students (much to my surprise) as an iron-clad declaration of classroom policy. As soon as I am seen to contradict the statement, students bring it to my attention, using a route in which their anonymity is guaranteed (such as the classroom CIQ, in which students respond weekly to a series of questions about their learning experience). I can then address this apparent inconsistency in class.

The second response is to be explicit about your commitments and convictions as a teacher in the course syllabus and then find some way of assessing once or twice a semester how consistently you are living them out. For example, every now and again a brief classroom research assessment instrument such as a "muddiest point" paper (where students are asked to write down the idea or activity that was the most confusing or unclear to them in class that day) might be devoted to this theme. So, in students' eyes, authentic teachers are those who walk the talk or practice what they preach.

Full Disclosure. Full disclosure refers to the teacher's regularly making public the criteria, expectations, agendas, and assumptions that guide her practice. Students know and expect us to have such an agenda and are usually skeptical of any statement to the contrary. After all, if we don't have criteria, expectations, agendas, and assumptions, what do we stand for, and why do we bother to show up for work? In Myles Horton's words, "There's no such thing as being a coordinator or facilitator, as if you don't know anything. What the hell are you around for, if you don't know anything? Just get out of the way and let somebody have the space that knows something, believes something" (Horton and Freire, 1990, p. 154). Unless you make your expectations, purposes, and criteria explicit, you will be perceived as holding them close to your chest, being secretive, and therefore not to be trusted.

It is interesting that even if students dislike a teacher's expectations and agendas, knowing clearly what they are because the teacher consistently makes them explicit builds trust with students. Students would much prefer to know what you stand for—even if they disagree with it—than to like you personally but be in the dark as to what it is you're expecting. So an important part of skillful teaching is to find a way to communicate regularly your criteria, assumptions, and purposes and then to keep checking in to make sure students understand them. At a minimum, your syllabus should contain a summary of your expectations and assumptions as well as an unequivocal statement of the criteria you are applying to judge student work. In students' eyes, where authentic teachers are concerned, what you see is what you get. There are no tricks played on them, no information held back, and no switching of the rules or direction midstream. As we shall see later in this chapter, this emphasis on early and complete disclosure as a condition of authenticity raises problems for teachers who work with students who are reluctant to engage with critical thinking or consider new and potentially threatening ideas or activities.

Responsiveness. Responsiveness is the dimension of authenticity stressed by Grimmet and Neufeld (1994) focusing on demonstrating clearly to students that you are teaching so as to be most helpful to them. Such clear student-centeredness has two elements. One is the teacher's constant attempt to show that she wants to know how and what students are learning, what inhibitors and enhancers to learning are present in her teaching, and what concerns students have about the course. The other is her public discussion with learners of how this knowledge affects her own teaching, including the extent to which some elements of the course can be negotiated. Responsiveness is not the same as capitulation or always bowing to the wishes of the majority. But it is taking the majority's wishes seriously enough to discuss with students why they cannot always be met, and then being ready to negotiate how particular learning tasks might be accomplished. To use an example from my own practice, I will not abandon my agenda regarding the teaching of critical thinking; that's why I'm in the classroom. But I do negotiate how students demonstrate such thinking if the assignments I have set are dissonant with their learning style, personality, or cultural formation.

Using a classroom assessment technique such as the CIQ is one important way to demonstrate responsiveness. The CIQ is a single-page form handed out to students once a week at the end of the last class I have with them that week. It comprises five questions, each of which asks students to write down some details about events or actions that happened in the class that week. Its purpose is not to ask students what they liked or didn't like about the class (though that information inevitably emerges). Instead it gets them to focus on specific events and actions that are engaging, distancing, confusing, or helpful. Having this highly concrete information about particular events and actions is much more useful than reading a general statement of preferences.

The CIQ takes about five minutes to complete; students are told *not* to put their name on the form. If nothing comes to mind as a response to a particular question, they are allowed to leave the space blank. They are also told that at the next class I will share the group's responses.

These are the five questions:

1. At what moment in class this week did you feel most engaged with what was happening?
2. At what moment in class this week were you most distanced from what was happening?
3. What action that anyone (teacher or student) took this week did you find most affirming or helpful?
4. What action that anyone took this week did you find most puzzling or confusing?

New Directions for Adult and Continuing Education • DOI: 10.1002/ace

5. What about the class this week surprised you the most? (This could be about your own reactions to what went on, something that someone did, or anything else that occurs.)

At the start of the first class of the next week, I spend five to ten minutes reporting back to students a summary of the chief themes that emerged in their responses. In my own teaching, the CIQ has been crucial in demonstrating responsiveness. Each week it affords a running commentary on how students are experiencing their learning and my teaching using words and examples that spring from students' own experience. In class, or online, I can talk about my reactions to these publicly disseminated student comments, say how they have challenged or confirmed my assumptions about the best way to teach the class, discuss any discrepancies that seem to be emerging between what I expect of learners and what they think I expect of them, and generally show that I take their opinions seriously enough to solicit them in the first place and then respond publicly. Learners tell me that in doing this I am showing them I can be trusted—that, in effect, I am an authentic teacher. Since being trustworthy is at the heart of authenticity, it's clear to me that being responsive to students in this way is crucial.

Personhood. Personhood is the perception students have that their teacher is a flesh-and-blood human being with a life and identity outside the classroom. Students recognize personhood when teachers move out from behind their formal identity and role description to allow aspects of their life and personality to be revealed in the classroom. Instead of being thought of as a relatively faceless institutional functionary, a teacher is now seen as a person moved by enthusiasms and dislikes. This is not to say, though, that a teacher should indiscriminately turn the classroom into a zone of personal confession. Personhood is more appropriately evident when teachers use autobiographical examples to illustrate concepts and theories they are trying to explain, when they talk about how they apply specific skills and insights taught in the classroom to their work outside, and when they share stories of how they dealt with the same fears and struggles that their students are currently facing. In Chapter Four, Kate Frego offers several practical examples of how teachers can establish a sense of self in the classroom.

When I first learned how important personhood was to students, I was reluctant to follow its tenets (I am English, after all). But because its presence seems to support students learning, I have tried to pay attention to this dynamic, particularly when teaching difficult material. One of my teaching preoccupations has been to introduce students to the body of work broadly known as critical social theory. My main concerns are to explain some of its central concepts in a way that is accessible but not overly simplistic, and to show how concepts such as alienation, hegemony, and commodification might illuminate students' lives. As I do this, I draw explicitly on how these ideas help me understand better what I have witnessed in workplace relationships and teaching practices over the years. I show how dominant ideology shapes my decisions as a teacher, how I unwittingly engage in self-surveillance and self-censorship, how hegemony causes me to conclude that I've been a good teacher only on those days when I come home completely exhausted, how repressive tolerance manifests itself in my attempts to open up a discussion or broaden the curriculum, or how automaton conformity frames my response to new practices or ideas. I use autobiographical examples, but only to help students understand core concepts in the course—not to tell entertaining stories for the sake of storytelling.

I also talk frequently about my own struggle in engaging with this tradition. I talk about how much time it takes me to read the texts, how I study the same sentence over and over again and still have no idea what it means, and how I frequently feel like an idiot compared to colleagues who seem fully comfortable with Gramsci, Althusser, Foucault, and Marcuse. Students consistently tell me what a shocking, though very welcome, revelation this is. They automatically assume (as I probably would in their place) that as the designated professor for the course I have got critical theory down. Interestingly, this admission does not seem to weaken my credibility. Instead, students seem relieved that someone who has studied this work for some time still feels like a novice. Again, my intent is that this autobiographical disclosure be done in the cause of supporting student learning, and that such disclosure increase my sense of personhood in learners' eyes.

Critiquing Authenticity Through the Lens of Power

If the teacher is honest with students, she must acknowledge that she has considerable power—the power to define curricula, set evaluative criteria, and then use these criteria to decide the worth of student work. Teachers cannot pretend this difference doesn't exist and simply be friends with students, though they can treat them respectfully and collegially. Sometimes the teacher's agenda is in direct conflict with that of the students; in such a case, it would be inauthentic for teachers to deny their identity by simply agreeing to do what students want. So being authentic involves staying true to one's agenda, remaining open and honest about it, and sometimes placing one's power behind it. This raises many complex questions. Can a teacher be authentic, yet practice ethical coercion? Can power be given away, or can it only be claimed? What power do students have to exercise in the classroom to keep the teacher honest? Are authenticity and authority compatible? In this section, I explore a central question: How do we exercise power in an ethical and responsible manner while being authentic? In so doing, I draw on several educators whose work is informed by critical traditions, particularly Ian Baptiste, bell hooks, and Herbert Marcuse.

The Grenadian-born adult educator Ian Baptiste (2000) is one who has considered at length questions of power and justifiable use of authority. He argues for an ethically grounded pedagogy of coercion in which adult educators help learners identify their "true enemies": those who "intend, *on principle,* to frustrate the goals of their opponent because their opponent's goals stand in opposition to theirs" (p. 29). To Baptiste, adult educators often function as persuaders and organizers, using justifiable coercion, but choose not to acknowledge doing so. In Baptiste's view, it is naïve, and empirically inaccurate, for adult educators to insist that their job is not to take sides, not to force an agenda on learners. Like it or not (and Baptiste believes most of us do not like to acknowledge this), adult educators cannot help but be directive in their actions, despite any avowal of neutrality or noninterference.

One of the most contentious aspects of Baptiste's writings is his insistence on the morality of coercion. Citing George S. Counts in his support, Baptiste believes that adult educators cannot avoid imposing their preferences and agendas on learners, and that in certain instances it is important to do so. Sometimes, in furtherance of a legitimate agenda or to stop perpetration of an illegitimate one, Baptiste argues that the adult educator must employ coercion. At other times, and for reasons that have to do with the adult educator's wish to stop any challenge to his authority, coercion is used but masked by a veneer of passive-aggressive, nondirective facilitation. We all know of situations in which we, or our colleagues, have said that anything goes, while concurrently making it very clear (often through subtle, nonverbal cues) that the "anything" concerned needs to reflect our own preferences. As I understand his position—advanced in a conference dialogue with me (Baptiste and Brookfield, 1997)—Baptiste would be skeptical of the notion that it is possible ever to be totally nondirective if the teacher has the judgmental power of the grade. This is a position Foucault (1980) would also underscore.

To support his case, Baptiste describes a situation in which he worked with a number of community groups on the south side of Chicago to assist them in reviving an area ravaged by pollution and migration. As the neutral and independent facilitator, he was supposed to stay free of forming alliances with any of the groups involved. Citing his liberal humanist sensibilities, he describes how, in trying to stay neutral, "I succeeded only in playing into the hands of the government officials (and their lackeys in the community). They played me like a fiddle, pretending in public to be conciliatory, but wheeling and dealing in private" (p. 47).

In hindsight, Baptiste argues, the experience taught him that in a situation where there is a clear imbalance of power, adult educators should take an uncompromising stand on the side of those they see as oppressed. An inevitable consequence of doing this is the necessity for them "to engage in some form of manipulation—some fencing, posturing, concealment, maneuvering, misinformation, and even all-out deception as the case demands" (2000, pp. 47–48). He points out that if an adult educator does admit that manipulation is sometimes justified, then an important learning task becomes researching and practicing how to improve one's manipulative capacities. Through studying ethically justified manipulation, adult edu-

New Directions for Adult and Continuing Education • DOI: 10.1002/ace

cators can "build a theory that can legitimize and guide our use of coercive restraint" (p. 49).

Baptiste's analysis raises some troubling issues for consideration of authenticity. In terms of his overall project, he is consistent in his pursuit of genuine (rather than counterfeit) participatory democracy, which seems to address the issue of congruence. It is around the issue of full disclosure that Baptiste breaks with students' perceptions of authenticity. Concealment and all-out deception are justified and inevitable in the hierarchical surroundings of a formal educational institution. Most teachers would accept that sometimes one must conceal one's intentions from the employer to ensure one is able to take risks, experiment, and generally do critical and challenging work. Much more contentious is the issue of whether it is ever authentic to conceal one's intentions from students. Yet for teachers trying to teach critically by raising uncomfortable and challenging viewpoints in class, making full disclosure of their intentions in advance can easily undercut the project.

So to be educationally effective—that is, to have students be ready to consider an alternative and troubling new perspective—a teacher may need to keep concealed the fact that students will be asked to do so until the teacher judges them to be at a point of learning readiness. The teacher may also wait to introduce these new perspectives until students' trust has been earned. Even if this is at an earlier stage in a course, it may be that when the teacher starts to introduce the new perspective or activity the acquired trust is then completely destroyed. In introducing previously unannounced and challenging material, a teacher may find that students see this as a fundamental and surprising change of direction. In this situation, such teachers are seen as acting in a way that contradicts their words, thus negatively affecting student perceptions of teachers' congruence (itself a prime indicator of authenticity). Introducing a new objective midway through the course also adversely affects students' perception of the teachers' full disclosure, another indicator of authenticity.

Yet requiring students to engage with new and challenging material is certainly justifiable. Indeed, one could argue, the most valuable learning that people experience often happens when they are forced to consider perspectives, information, and realities they would prefer to avoid. This illustrates a contradictory dynamic: attending assiduously to creating conditions of authenticity through making full disclosure in advance of one's agenda is opposed by the equally justifiable need to conceal significant information about the learning agenda.

There is another contradictory dynamic at play here as far as authenticity is concerned. As we have seen, a prime indicator of authenticity is the teacher's clear responsiveness to learners' concerns. Even so, students' long-term intellectual development sometimes requires that we *not* do what they ask, thereby appearing to be unresponsive to their wishes. For example, I know that when I'm trying to get students to think critically I must sometimes refuse to comply with requests made in class or on the CIQs to tell them the correct view to hold on an issue, or the right assumption to follow in a certain situation. My refusal to tell them the proper opinion or what they should be thinking—in effect, to refuse to comply with their request for the right answer—appears to contradict the condition of responsiveness. If this is added to my insistence that students engage with ideas they would rather avoid, or that they undertake activities they don't see as relevant, I appear to be the most inauthentic of teachers, if authenticity is judged by responding promptly to learners' expressed needs. In all these situations, I would argue it is educationally important that I use my power as a teacher to do things I determine will enhance students' long-term intellectual development, even if in the short run it makes me appear inauthentic.

One theorist who has explored this dynamic is Marcuse (1965). He argues it is educationally crucial that learners be exposed to alternative, often dissenting, ideologies and perspectives even though they do not see the necessity for doing so. To him, this is the practice of liberating rather than repressive tolerance. Marcuse argues that without knowing of the full range of options, viewpoints, or ideologies surrounding an issue, students cannot make a truly informed judgment as to which directions they wish to explore more deeply. If students have the choice, Marcuse argues, they usually choose curricular directions and learning activities that are familiar and comfortable. These directions and activities are ones that have, in effect, been ideologically predetermined by students' previous histories and experiences. Students choose learning projects that support and confirm prevailing ideology and steer clear of anything they sense is "deviant" or "in left field." The teacher's duty (according to Marcuse) is to spend a considerable amount of time exposing students *only* to ideas and activities they would otherwise have avoided. This is the only way a teacher can ensure students will be availed of the full range of perspectives and opinions that exist on any issue.

The problem regarding authenticity is that to teach against students' wishes appears to them as being unresponsive to their concerns and desires—and responsiveness to student concerns is one of the four main indicators of authenticity learners identify. If the teacher stands by the conviction that students need to learn about alternative theories and dissenting viewpoints that they are uninterested in, then the teacher risks being perceived by the students as rigid and authoritarian. As I pointed out earlier, this is why we must never confuse responsiveness with capitulation to majority wishes or always doing what students say they want. Instead, we must understand responsiveness as fully addressing learners' concerns and questions, even if this means rejustifying why we can't do what they say they want us to do. In this situation, being responsive is to explain as fully and convincingly as possible why you believe sticking to your agenda is in their best long-term interests, even if they violently disagree with your position. Of course, this is much easier said than done, and (as we shall see in a moment) there are often strong personal, institutional, and professional pressures to teach in a way that pleases students.

New Directions for Adult and Continuing Education • DOI: 10.1002/ace

The African American feminist bell hooks has much to say on how the indicator of responsiveness brushes up against the inevitability of teacher power. She views the feminist classroom as an arena of struggle distinguished by striving for a union of theory and practice. One of the most striking elements in her analysis of such a classroom is her emphasis on how the exercise of teacher power is often unavoidably—even necessarily—confrontational. In her judgment, the teacher's position "is a position of power over others" with the resultant power open to being used "in ways that diminish or in ways that enrich" (hooks, 1989, p. 52). To emphasize the commitment students should have to the learning of others, she takes attendance, a practice reminiscent of elementary school for many skeptical adult students. To underscore the importance of attendance, she lets students know that poor attendance negatively affects their grade. She requires that everyone participate in class discussion, often by reading aloud paragraphs they have already written.

Such practices inevitably lead to negatively critical comments by students, a fact that she admits has been difficult for her to accept. Because "many students find this pedagogy difficult, frightening, and very demanding" (hooks, 1994, p. 53) teachers who use it will be resisted, even disliked. Students may also elect not to take their courses. This is why hooks insists that the humanistic emphasis on having students perceive the classroom as a safe, positive, and congenial environment for learning is not a good criterion to use in assessing teacher competence. There are professional consequences to receiving poor evaluations, such as being denied reappointment, losing merit pay, or having promotion or tenure refused. If this is the case, then the institutional pressure is on for teachers to work in a way students find pleasing and familiar. If, on the other hand, teachers insist on sticking to their guns and requiring students to engage with activities and ideas they would much rather avoid, they will be seen as contradicting the criterion of responsiveness. In many students' eyes, a responsive teacher is one who, faced with student resistance to a particular learning activity (such as engaging in a role play to illustrate the ubiquitous reality of racism, patriarchy, or class bias), agrees to do something more to the students' liking. Again, this difficulty is resolved if we understand responsiveness as being different from capitulation, and as involving responding to student criticisms and requests by explaining as fully as possible why they cannot be met. But in the short run, this is often a hard stance for a teacher to take.

I believe it is possible, though difficult, for teachers to maintain a relatively congenial state of tension between being authentic and exercising power. In such a situation, students can disagree with a teacher's exercise of power while possessing full knowledge of why she is insisting on her agenda and refusing to comply with student demands. At the heart of authenticity is the matter of trust, and part of being trustworthy is presenting as honest a picture as possible of one's agenda and convictions. Although in the short run students might disagree strongly with a teacher's direction, they will

New Directions for Adult and Continuing Education • DOI: 10.1002/ace

believe the teacher to be honest if the teacher makes full disclosure as to why the direction is being pursued in the face of student dissent.

The more fundamental and essentially irresolvable contradiction in being both authentic and true to one's agenda as a critical teacher arises when one is trying to bring students round to the point of learning readiness where they are willing to consider ideas and activities that they would otherwise have ignored or derided. In such a situation, authenticity (if interpreted as full disclosure) and teaching for intellectual development (if understood as requiring students to stretch themselves in ways they would not themselves have chosen) may be directly at odds. After thirty-five years of trying to resolve this contradiction, I have realized it is irresolvable—one of the ontological and practical contradictions that we have to live with even as our institutions pretend that teaching (defined as sequenced, orderly managing of student learning to achieve predetermined outcomes) is always free of ambiguity.

References

Baptiste, I. "Beyond Reason and Personal Integrity: Toward a Pedagogy of Coercive Restraint." *Canadian Journal for the Study of Adult Education,* 2000, *14*(1), 27–50.

Baptiste, I., and Brookfield, S. D. "'Your So-Called Democracy Is Hypocritical Because You Can Always Fail Us': Learning and Living Democratic Contradictions in Graduate Adult Education." In P. Armstrong (ed.), *Crossing Borders, Breaking Boundaries: Research in the Education of Adults.* London: University of London, 1997.

Cranton, P. *Becoming an Authentic Teacher in Higher Education.* Malabar, Fla.: Krieger, 2001.

Cranton, P., and Carusetta, E. "Perspectives on Authenticity in Teaching." *Adult Education Quarterly,* 2004, *55*(1), 5–22.

Foucault, M. *Power/Knowledge: Selected Interviews and Other Writings, 1972–1977.* New York: Pantheon Books, 1980.

Grimmet, P. P., and Neufeld, J. (eds.). *Teacher Development and the Struggle for Authenticity: Professional Growth and Restructuring in the Context of Change.* New York: Teachers College Press, 1994.

hooks, b. *Talking Back: Thinking Feminist, Thinking Black.* Boston: South End Press, 1989.

hooks, b. *Teaching to Transgress: Education as the Practice of Freedom.* New York: Routledge, 1994.

Horton, M., and Freire, P. *We Make the Road by Walking: Conversations on Education and Social Change.* Philadelphia: Temple University Press, 1990.

Marcuse, H. "Repressive Tolerance." In R. P. Wolff, B. Moore, and H. Marcuse (eds.), *A Critique of Pure Tolerance.* Boston: Beacon Press, 1965.

Palmer, P. J. *The Courage to Teach: Exploring the Inner Landscape of a Teacher's Life.* San Francisco: Jossey-Bass, 1997.

STEPHEN D. BROOKFIELD *is distinguished university professor at the University of St. Thomas in Minneapolis-St. Paul.*

New Directions for Adult and Continuing Education • DOI: 10.1002/ace

2

This chapter examines the myth of "women's learning" and suggests how to develop a teaching practice that is authentic, open to difference, and attentive to power.

Women, Knowing, and Authenticity: Living with Contradictions

Leona M. English

Do women know and learn differently from men, and is this difference accurately reflected in our practice? If we are to believe much of the writing on women and learning, women are just a bit more relational (Fletcher, 1998), just a bit more caring and connected (Giraldo, 2002), just a bit warmer and friendlier than their male friends and colleagues, not to mention more inclusive (MacKeracher, 2004).

Despite the critiques of essentialist thinking about women (Hayes, Flannery, Brooks, Tisdell, and Hugo, 2000), there is still a subtext in much of this writing on women and learning: that adult educators ought to do more to reach women and help them feel included, nurtured, and loved, as well as let them have a voice. As helpful as this literature may have been, I argue in this chapter that this subtext can lead to a simplified portrayal of authentic women as homogeneous, uncomplicated, and harmonious, and it can also lead to some misguided and simplistic attempts to plan learning experiences for them. As a response, this chapter presents a more complicated way to look at women's learning that resists these stereotypes. I examine some of the contradictions inherent in how women came to be seen as One Woman and suggest how we might approach their learning more authentically. I begin with positioning myself in the research, move to exploring the current state and influences on the field of women's learning, then take up current theories of feminism and post-foundationalism, and finally discuss the implications of these new advances for teaching and learning.

NEW DIRECTIONS FOR ADULT AND CONTINUING EDUCATION, no. 111, Fall 2006 © 2006 Wiley Periodicals, Inc.
Published online in Wiley InterScience (www.interscience.wiley.com) • DOI: 10.1002/ace.224

Positioning Myself

I am a female academic who started my professional life as a high school teacher of English. I have taught in education, adult education, and ministerial programs. Most of my students have been women; about 85 percent of my current students in a master of adult education program are women. Over the years, I have been involved in women's organizations, as well as a variety of activist and social justice causes. I also socialize with women in organizations and female-dominated professions, and what I have experienced about authentic women and learning is more complicated than what any of the stereotypes allow. Power relationships, resistance, and desire are all part of this learning, though little of it is accounted for in the standard adult learning literature.

What Does It Mean to Be Authentic?

Authenticity is a much contested subject in this age of postmodern sensibility, when anything that speaks of a core self or a cause is called into question. Authenticity seems to be a modernistic sentiment that acknowledges a pure self, one that can be uncovered, examined, and developed if only we are critically reflective enough. I argue in this chapter that we can have an authentic vision of women and learning that is neither modern nor rooted in essentialisms, unitary selves, or static notions of women's learning. In deconstructing authenticity, my commitment is to a simultaneous reconstruction of knowledge. Therefore I use the term *authentic* in a critical sense, holding on to its connotation of genuineness or helpfulness and rejecting the attendant notion that there is a static, immutable self that cannot change. I offer ideas and practices on women and learning that are open, challenging, fluid, and exploratory, yet authentic. I agree wholeheartedly with Dei's comment (2002) that "any claim to authenticity as possessing authority or authoritative voice" must be open to critique (p. 9).

So Where Did Our Notion of Women's Learning Come From?

So ingrained are adult education's notions of women and learning that we have an entire vocabulary dedicated to them—feminist epistemology, feminist pedagogue, feminist pedagogy, circle pedagogy, voice, connected learning—much of it derived from *Women's Ways of Knowing* (Belenky, Clinchy, Goldberger, and Tarule, 1986). Building on the work of Carol Gilligan (1982), Belenky and colleagues developed the notion that women had particular ways of learning that accentuated the connectedness of their lives and their learning processes. They advanced the notion, perhaps inadvertently, that women were different from men when they learned (women are connected knowers and men are separate knowers) and that women were

New Directions for Adult and Continuing Education • DOI: 10.1002/ace

essentially kinder and gentler learners who preferred experience-based knowing. It became a given that our pedagogical practices ought to reflect their insights.

In an attempt to redress the inequities of women's learning practices, there was a wave of attention in the 1980s and 1990s to women and their learning, which might have contributed unwittingly to further essentialization of women and learning (Goldberger, Tarule, Clinchy, and Belenky, 1996). Within adult learning theory, writers such as Hayes, Flannery, Brooks, Tisdell, and Hugo (2000), for instance, have addressed some of this essentialism, but it lingers to some degree in our practice and our thinking. For example, Jane Hugo (2001) says, "women have a particular way of learning and knowing that is grounded in making connections" (p. 90). Although she is blatant in her assertion, most adult education writing is more subversive and perpetuates these ideas in a more careful and nuanced way, which is no less essentialized and no less harmful. Flannery and Hayes (2001), for instance, say there is as much difference among women as there is commonality, and that adult learning is "multidimensional, fluid and diverse" (p. 39). Still, there are few who would argue that adult educators need to attend to women's ways of learning.

How did we get this way? The theory of social constructionism, which has helped us understand that men and women are often socialized in their own way and that difference affects how we learn, has also led to some of this thinking that women's learning needs to be warm and fuzzy. Social constructivism and all that has grown out of it has its place. We are thankful attention was paid to women; we are thankful that the social roles in constructing identity have been given due attention; we are glad that women's ways of knowing have been recognized. But there is no doubt that this emphasis on the social construction of learning, womanhood, and other gendered identities is circular. Women are socialized as warm and caring, so we must teach them in ways that are warm and caring; women are warm and caring because they have been socialized as such, and so on. We need new lenses and a fresh perspective that is more complex, has fewer binaries or polarities, and is inevitably more challenging.

Clearly, the women's learning conversation has run its course. We must reach beyond the humanistic and psychologized literature, from which most of this thinking has derived, and allow newer critical feminist and postmodern studies to come into play.

What the Research Is Telling Us

Adult education has not kept pace with research in the social sciences. Here is a snippet of what we are learning from feminist and postfoundational studies.

Learning from Feminist Studies. The predominant view of women's learning operative within adult education was influenced by the first and second waves of feminism (1960s and 1990s). This feminized and separate view served a useful purpose by drawing attention to women's causes and women's exclusion in all realms of education and learning. This separate view came to be replaced in the 1990s with third-wave feminism, which was concerned with difference; inclusiveness; and issues of gender, race, and class (Starr, 2000). This third wave honors the fact that there is no one way of being a feminist or a woman. It recognizes the interrelatedness of feminism to all areas of study and to global concerns, moving it out of individualized and psychologized preoccupations such as self-development, personal growth, and women's liberation. The third wave allows us to honor the female student who likes studying independently, attending lectures, taking notes, and doing papers. Unfortunately, adult education's understanding of women's learning did not keep pace with these trends in third-wave feminism.

In cognate disciplines and fields, the changes were more rapid. Witness the transition of women's studies programs in universities that blossomed in the 1970s and 1980s, replaced in the 1990s by gender studies, which included concern for males and females, and now in the new millennium being replaced by interdisciplinary programs in environmental, social justice, and globalization studies, which see feminist issues as part of a larger global discourse. Nevertheless, adult education continues to segregate women's learning as a separate sphere, further isolating women from authentic relationship and knowing. One could argue that we have so many groups of all kinds that all of us, including men, are isolated in some way. The focus here remains on how women have become essentialized and stereotyped.

Emergent Understandings in Postfoundationalism. Along with the advances in feminism, the postfoundational areas of research (which include postmodernism, postfeminism, postcolonialism, and poststructuralism) are unsettling sureties such as women's ways of knowing and men's ways of knowing (English, 2005; Tisdell, 1998). They challenge modernity's promises of quick cures, unbridled growth, and fixity of any sort. Postfoundationalism plays with the notion of difference and does not seek solid identity, essentialism, or core self. Postmodernity, in particular, challenges the grand narrative that there is one way to teach, and that personal growth is desirable and achievable (Harvey, 1989). Postmodernity advocates more fluid understanding of teaching and learning. This implies allowing for difference among and between men and women, of cultivating uncertainty instead of surety in areas such as women's learning, and of questioning and critiquing taken-for-granted notions such as that women learn one way or another. This challenges any notion of a unitary self ("the learner," or "the educator," or "the woman as learner or teacher"). It helps us see identity as shifting, nonunified, and constantly changing, all of which makes attempts to codify teaching practices for any group moot.

Poststructuralism draws attention to the interrelationships among power, knowledge, and discourse, making gender very much a subcategory of much larger issues. Rather than focus on a list of gender differences, poststructuralism "informs an analysis of how women experience and express shifting identities that reflect multiple social influences" (Hayes, Flannery, Brooks, Tisdell, and Hugo, 2000; Ryan, 2001). Poststructuralism is a body of theory that acts as a resistance to the solidity and surety of the structural movement, or the belief that underneath everything there is a complex and ordered system. The branch of poststructuralism that is informed by Foucault is more interested in how power operates among and between all learners and educators. Power is seen to be disciplinary and exercised in all our interchanges and relationships (Foucault, 1980). This draws attention to how we govern ourselves in a learning situation. The notion that women are always acted on or subservient in the learning situation, for instance, is challenged by the concept of power circulating everywhere and being exercised by everyone. In the traditional classroom, the educator might look as if he or she holds all the power but the learner fills out the anonymous evaluation, can resist by tuning out, and can choose to do or not do the work (or not to sign up). The learner exercises power through resistance, which can be quite productive.

Despite the fact that we think we are "empowering" women in our classes, we may in fact be producing self-discipline, or having women "control" themselves in fear of what others (we included) might say or think. Regardless if others are watching or not, there is a tendency to think they are, and to act in a way that conforms to what the watcher might want. This self-discipline of saying the right thing or conforming to unwritten group rules is a way in which we exercise power over our own body and action. We use the feminist pedagogical practice of putting women in groups, and the result is often that learners think they are being watched and so act as if they were. They try to think of brilliant things to say and do. This is the effect of the grouping exercise that was supposedly empowering. This understanding of power helps us see that even the most egalitarian of teaching spaces is always power-ridden.

All of these approaches challenge how women are as learners. They help us speak to what is authentic in the teaching and learning process for both males and females. These social science theories have ramifications for how we teach, learn, and do research with all learners. They may constitute a way for us to trust who we are as educators, as those who respond to the learner in shifting and uncertain times. One of our challenges as educators is to be open to change; accepting of difference; and receptive to new and fluid notions of self, learning, and teaching. Our task is not to hold on to the old, static notions of learners and educators; rather, we need to become colearners and coexplorers. This, of course, does not mean that we forgo common sense; rather, it means we remain open to difference. In these changing times, we may need to take better care of ourselves, know ourselves better, and develop life-sustaining practices (such as exercise) that allow us to be healthy and open. Educators need to be comfortable with some degree of uncertainty and ambiguity in practice; only in strengthening our inner resources can we be prepared for change.

New Directions for Adult and Continuing Education • DOI: 10.1002/ace

Teaching and Learning Implications

How women come to know is as varied as women's being. Feminist and post-foundational work helps us see how we need to valorize the unknowable, the different, and the changing dimensions of learning and learners. We have to read how we do our work a bit differently. Rather than looking at how women create and learn, we need to change our focus so it is more open.

Authentic teaching is more complex than isolating women or teaching them in fixed "feminist ways." Not all learners want to sit in circles or write about their feelings; we need to see the complexity of teaching not just for women but for men as well. Here are some practical ideas that arise from this research.

Geometric Formations Include More Than Circles. At the heart of feminist teaching practice or pedagogy is the notion that if we sit in a circle and share our thoughts, experiences, and feelings, we will feel more included because we have heard our own voices. Our poststructural theories (Brookfield, 2001; Plumb, 1995) show us that the circle and related kinds of practices are examples of where power operates at its most effective level. The circle, like any teaching formation in the classroom, can produce effects such as resistance to speaking, fabrication of experience, or anxiety. Rather than create inclusive learning communities, circles can indeed be problematic. As educators, we need to think of how these teaching circumstances and configurations affect learners (and us), be attentive to the power circulating in them, and vary them (Foertsch, 2000). This might mean that we use a variety of seating arrangements or ask learners how they would like to sit; and even when we do use circles, we can allow silence and a variety of ways of participating, including optional writing assignments or projects in lieu of dialogue.

Pedagogical Practices Need to Welcome All. We need to understand that a number of the traditional adult education and learning practices have not been freeing for everyone. Learners have not always welcomed the self-revealing practices of reflexive inquiry or life narrative, which poststructuralism might call technologies of disciplinary power. Many of these teaching practices have in fact produced effects such as women's self–regulation or careful crafting of their narratives to please the teacher. Pedagogical practice needs to be more open and inclusive of difference. Not all women go to class to build community, share feelings, or socialize. Some go to hear the teacher speak, especially if they have little background in a topic or feel a need for factual information on some topic on the course syllabus. Educators might also think about creating options in assignments so that learners can decide how much personal information it is appropriate to reveal, or conducting individualized needs assessments.

Voice Can Be Heard or Silent. I have heard it said jokingly among adult educators that they would not take their own classes because they involved too much group work, talking, and sharing. This is sobering indeed. Feminist and postfoundational theories underscore the variety and difference in and among learners. The grand narrative of democracy and its implicit assumptions that everyone is free and wants to speak out is challenged by postmodernity. Learners have a variety of ways of speaking, only some of which include voice (dialogue, writing, acting). Designated grades for participation, for instance, may intimidate some learners; educators might think of either inviting learners to assign their own grades for participation or encouraging them to participate in defining what active participation might mean in their context. For some learners, this could entail more writing, small group work, one-on-one meetings with peers, or dialogue with the educator. The assumption that everyone wants to contribute verbally or that verbal participation is a must for women needs to be continuously challenged in practice.

Contextualize Women and Learning. We need to set the discussion of women and learning within a larger discussion that embraces issues such as feminism, globalization, and learning. This broader discussion helps move our emphasis from a list of gender differences to larger issues and helps educators and learners put more emphasis on the meaning being created and less on the individual person. Just as women's studies programs have made the transition to more complex integrated studies programs, so too must our understanding of women and learning make the transition to a larger discussion of the purpose of learning and the global resources for learning. Educators might do this by changing courses from women and learning to gender studies, or to issues in learning, or globalization studies. In this way, women's issues in learning can be seen as part of a larger social construct and as including a broader discussion of categories of analysis such as class, race, and ableness. Retaining an isolated focus on women further marginalizes women and creates the illusion that women's issues are somehow divorced from the social and cultural arena in which they arise.

Attend to Issues of Power. Sometimes, even women are not nice. We need to resist any essentialism and challenge students to examine their own assumptions about pedagogy, gender, and learning. It is important to openly discuss issues of power and how it works through the classroom and learning environment—look at issues of gender and at issues of how power flows in relationships, who talks, when, and why. We also need to be open to seeing how learners or adult educators exercise silence or self-discipline in order to control others or the situation. Learners might be encouraged to study theories of power and complete ethnographic studies of power in their workplace or class. By actually cultivating faculties of perception about power, learners become more aware of and sensitive to how power operates. They learn to question teaching and learning practices and engage in continuous critique of power in their own practice.

As Foertsch (2000) reminds us, we need to be attentive to and make clear how we, as educators, use feminist practices deliberately to "facilitate" good teaching. In using circles, student-led presentations, collective pronouns (the use of "we" to suggest that everyone in the class worked together to create a democratic space), and small group learning, we are often in a position of overseeing the learning, "guiding" students to the right answers, and showing them the correct path. Though defensible as our responsibility to be good educators, these practices are power-laden and need to be recognized as such.

Pay Attention to the Resistances. Whether the learners be male or female, wherever there is teaching and learning there is power, wherever there is power, there is resistance. We need to pay attention to how resistance happens in all groups of learners, no matter the group composition. So, for instance, in the humanistic practice of self-directed learning we should pay attention to how learners resist (as in continuing to look for teacher direction) and to how our technologies or practices of power produce a desire in students to learn more, avoid others, or choose not to follow directions (see Chapman, 2003). Educators might trace how power runs through the situation by keeping their own professional journal, videotaping (with permission) their own classes, and inviting feedback from learners and colleagues.

Final Thoughts on Teaching and Learning

Of course, in searching for women's learning and ways in which to be part of it, adult educators are left with contradictions that must be analyzed and lived with. Can we ever be authentic if we are as moving, shifting, and fluid as the social science theories seem to suggest is our reality? Is women's learning really as malleable as they suggest? Being authentic in our teaching may mean we need to acknowledge the complexity of women's learning and explore further the dynamics of women, power, and learning.

Marshall Berman may be right, quoting from Keats that there is possibility in "negative capability" or the "power to live with . . . inner contradictions" (Berman, 2004, p. 85). To be authentic may indeed be to hold all these contradictions at once—to be open to the facts that some learners want direction and some do not and that power shifts are always affecting authenticity. To be authentic, educators must continue developing capacities of reflexivity, questioning their own practice, and developing new ways of communicating with learners. The ability to be open and inviting of change may indeed bring freedom to the learner and the educator.

References

Belenky, M. F., Clinchy, B., Goldberger, N., and Tarule, J. *Women's Ways of Knowing.* New York: Basic Books, 1986.
Berman, M. "Israel: No Souvenirs." *Dissent,* 2004, 51(3), 82–86.
Brookfield, S. D. "Unmasking Power: Foucault and Adult Learning." *Canadian Journal for the Study of Adult Education,* 2001, 15(1), 1–23.
Chapman, V.-L. "On 'Knowing One's Self' Selfwriting, Power, and Ethical Practice:

Reflections from an Adult Educator." *Studies in the Education of Adults,* 2003, *35*(1), 35–53.

Dei, G. J. "Spiritual Knowing and Transformative Learning." NALL working paper, no. 59, 2002 [http://www.oise.utoronto.ca/depts/sese/csew/nall/res/59GeorgeDei.pdf]; retrieved Feb. 28, 2006.

English, L. M. "Poststructuralism." In L. M. English (ed.), *International Encyclopedia of Adult Education.* London: Palgrave, 2005.

Flannery, D. D., and Hayes, E. "Challenging Adult Learning: A Feminist Perspective." In P. V. Sheared and P. A. Sissel (eds.), *Making Space: Merging Theory and Practice in Adult Education.* Westport, Conn.: Bergin and Garvey, 2001.

Fletcher, J. K. "Relational Practice: A Feminist Reconstruction of Work." *Journal of Management Inquiry,* 1998, *7*(2), 163–186.

Foertsch, J. "The Circle of Learners in a Vicious Circle: Derrida, Foucault, and Feminist Pedagogic Practice." *College Literature,* 2000, *27*(3), 111–129.

Foucault, M. *Power/Knowledge: Selected Interviews and Other Writings 1972–1977.* New York: Pantheon, 1980.

Gilligan, C. *In a Different Voice.* Cambridge, Mass.: Harvard University Press, 1982.

Giraldo, M. A. "Contributions for Strengthening and Promoting Processes of Social Organization into Organizations with Political Processes." In C. O. Houle Scholars in Adult and Continuing Education Program, *Global Research Perspectives,* Vol. 2. Compiled by R. M. Cervero, B. C. Courtenay, and C. H. Monaghan, May 2002 (N. Ashcraft, trans.). (ED 470936)

Goldberger, N., Tarule, J., Clinchy, B., and Belenky, M. (eds.). *Knowledge, Difference and Power: Essays Inspired by Women's Ways of Knowing.* New York: Basic Books, 1996.

Harvey, D. *The Condition of Postmodernity.* Cambridge, Mass.: Blackwell, 1989.

Hayes, E., Flannery, D., Brooks, A., Tisdell, E., and Hugo, J. *Women as Learners.* San Francisco: Jossey-Bass, 2000.

Hugo, J. "Creating an Intellectual Basis for Friendship: Practice and Politics in a White, Women's Study Group." In V. Sheared and P. A. Sissel (eds.), *Making Space: Merging Theory and Practice in Adult Education.* Westport, Conn.: Bergin and Garvey, 2001.

MacKeracher, D. *Making Sense of Adult Learning* (2nd ed.). Toronto: University of Toronto Press, 2004.

Plumb, D. "Declining Opportunities: Adult Education, Culture, and Postmodernity." In M. R. Welton (ed.), *In Defense of the Lifeworld: Critical Perspectives on Adult Learning.* Albany: State University of New York Press, 1995.

Ryan, A. B. *Feminist Ways of Knowing: Towards Theorising the Person for Radical Adult Education.* Leicester, UK: National Institute of Adult Continuing Education, 2001.

Starr, C. "Third Wave Feminism." In L. Code (ed.), *Encyclopedia of Feminist Theories.* New York: Routledge, 2000.

Tisdell, E. J. "Poststructural Feminist Pedagogies: The Possibilities and Limitations of Feminist Emancipatory Adult Learning Theory and Practice." *Adult Education Quarterly,* 1998, *48*(3), 139–156.

LEONA M. ENGLISH *is an associate professor of adult education at St. Francis Xavier University in Nova Scotia.*

3

*Central to development of authenticity in teaching is self-
understanding and self-awareness. Using a Jungian perspective, the author suggests that the imaginative
dimensions of the self play a critical role in our journey
and experience as teachers, and in developing self-
awareness and authenticity in our teaching.*

Authenticity and Imagination

John M. Dirkx

As he turned off the lights and closed the classroom door, Francis felt
tired—drained and still reeling from what seemed like a stinging attack by
one of the students in his graduate class on adult development.

Although most of the evening had seemed to go well, near the end of
the session Sara, a white woman her in early thirties and a practicing nurse,
challenged his claim that understanding developmental changes in adult-
hood was important to helping adults learn. At first, her questions seemed
thoughtful but respectful. Then her tone became more intense and even bit-
ing, as if she were cross-examining his whole perspective on adult learning.
Several students seemed to support Sara's openly critical perspective; others defended the value of developmental theory in working with adult learners, and the discussion became increasingly emotional and heated. Francis
listened carefully to Sara's initial questions and the ensuing discussion. But
then, as he sought to explain and elaborate his position, he felt himself grow
defensive and red-faced. As the discussion became more intense, Francis felt
helpless to constructively intervene. As the end of the session approached,
a couple of younger men cracked jokes about growing old prematurely from
overanalysis, and this seemed to dissipate the building tension. Students
packed up their books and headed out, wishing Francis and one another a
good night. A few stayed to talk for a few minutes about the questions that
were raised; they commented, as they were leaving, that they felt it was a
good session.

Despite these comments, Francis felt weary. As he walked alone down
the dimly lit, deserted hallway ninety minutes from midnight, feelings of
being incompetent and unloved by his students washed over him—feelings,

NEW DIRECTIONS FOR ADULT AND CONTINUING EDUCATION, no. 111, Fall 2006 © 2006 Wiley Periodicals, Inc.
Published online in Wiley InterScience (www.interscience.wiley.com) • DOI: 10.1002/ace.225

he observed, that haunted him through most of his twenty years of teaching. It would be better in the morning, he mused, when he was less tired. I'm a good teacher, he thought to himself, who cares about my students. Still, he wondered why he felt these things. Why were they so important to him? What effect did they have on his ability to help his students learn?

Through this experience, Francis reveals the heart of teaching, how the activities of teaching are intimately bound up with the self of the teacher in relationship with his or her students. Like other forms of practice in the so-called helping professions, the quality of one's teaching is deeply intertwined with and mediated through the interactions and relationships that teachers establish with their students. Because it is fundamentally rooted in relationship, teaching practice is inherently affect- and emotion-laden. As in other helping professions, this emotional dimension can—and often does—subvert our best intentions, leading us at times to act more in our own self-interest than for those to whom we are providing service. For example, studies demonstrate how the practice of medicine is embedded in powerful emotions that influence the quality of care provided, as well as the sense of self that doctors experience in their work (Balint, 1964; Mizrahi, 1986; West, 2001). Parker Palmer (1998) demonstrates how even the most committed teachers can, after years of dedicated service, lose heart.

Notwithstanding recent emphasis on the importance of learner- and learning-centered teaching, we in higher and adult education often minimize or ignore the potent emotional context in which our work is embedded. As with Francis, experience of these emotions may distract us from being as learning-centered as we would like. Rather than really listening to learners, we often assume they do not understand what we try to say, and we become preoccupied with clarifying or defending our position. At various times in my own teaching, I have felt annoyed, irritated, angered, or even threatened by what students said in class. In these moments, it sometimes seems as if my actions and responses are more defensive than helpful, intended more to protect or nurture a fantasized image of myself and of what my students must think of me. I may react with a flurry of perspectives, analysis, citations, and examples, seemingly drowning the unsuspecting student in a sea of information. Other teachers may respond differently, as with flooding students with niceness. In the quiet recesses of our fantasy, we might imagine such students as immature, wanting to show off to the rest of their peers, seeing if they can snag us in a logic trap or wanting all our attention. These images can serve to further intensify and justify our emotional reaction. But they may also suggest that we are being less than authentic in our teaching.

The emotional context of teaching—its heart and soul—metaphorically directs us to consideration of the extrarational and imaginative dimensions of teaching. In this chapter, I explore these dimensions and how we might engage the imagination to help us develop a deeper understanding and appreciation of the self as teacher and what it means to act authentically within one's teaching. In embracing the role of the imagination in self-knowing, I seek a way of understanding how the *psyche* or soul speaks to us through our teaching, and how it uses emotion-laden images and fantasies. Following the lead of others, I refer to this perspective as the *imaginal method* (Watkins, 2000) or, more simply, soul work (Moore, 1992; Peppers and Briskin, 2000).

To develop this perspective, I first review the connection between the self and authenticity in teaching, and pathways to the self and self-understanding as methods for fostering authenticity. I then focus more specifically on how we might work with the imagination to develop and sustain authenticity within our teaching.

The Self and Authenticity in Teaching

The central thesis of this chapter is that self-knowledge is at the heart of authenticity and represents the core of authentic teaching (Palmer, 1998). The idea of "authenticity" (Cranton, 2001; Cranton and Carusetta, 2004) places emphasis on qualities of the teacher as a person, on the nature of the self. It expresses the genuine self within a community, consistency between values and actions, relationships with others, and maintaining a critical perspective (Cranton and Carusetta, 2004). "To be able to express the genuine self," Cranton and Carusetta (2004) suggest, "people need to know who the self is" (p. 7). In the opening vignette, Francis seems to be struggling with the authenticity of his teaching. He is concerned not with his expertise in the subject matter and his ability to facilitate discussion but rather his sense of self that appears to be at the core of his deeply felt concerns about teaching.

The effectiveness of teaching is mediated largely through one's knowledge of the subject matter, related experience and background, skill in being able to render the subject matter accessible and meaningful to learners, ability to listen to students, and capacity to understand and appropriately respond to their struggle to learn. In varying degrees, the self of the teacher is called on to help render the subject matter understandable by and meaningful to adult learners.

Unlike programmed instruction machines, however, the teacher's actions arise from her or his own sense of self. Palmer (1998) argues that technique is what people use until the teacher shows up. As his work suggests, the self of the teacher is at the heart of good teaching. Francis's experience reflects both the actions of teaching that evening and his sense of himself as a person and a teacher, which has evolved over many years of experience. Teaching with a sense of authenticity reflects a profound sense of self-awareness and self-understanding. It draws our attention to the character of the teacher, its importance in the overall quality of our relationships with learners, and the effectiveness of learning experiences that we as teachers plan and facilitate. Teaching with authenticity mirrors Palmer's call (1998) for the teacher's need to attend to the soul as well as the role of teaching, to how one's identity and integrity are or are not manifest within the work of teaching.

Scholars in adult and higher education have begun to lay out various pathways to the development of self-understanding and authenticity in teaching. In considering the role of self-knowledge in teaching, however, we face many difficult questions. Who and what is the "self" from which the teaching act derives? How do we understand the development and fostering of this self-knowing that is at the center of authenticity in teaching? What does it mean to "know who the self is" (Cranton and Carusetta, 2004, p. 7)? How is that self-knowledge acquired or developed?

Consistent with a socially constructed, multiplistic, and nonunitary view of the self (Clark and Dirkx, 1999), many scholars stress the use of critical reflective processes to develop a deeper understanding and awareness of who and what the self is (Brookfield, 1995; Mezirow, 2000), and how our sociocultural contexts influence and shape who we take that self to be (Cranton, 2001; Cranton and Carusetta, 2004). Critical reflection represents a potentially powerful way to help teachers identify, critique, and possible modify existing assumptions and perspectives about themselves as teachers and the teaching-learning process.

Critical reflection suggests a process centered in one's ego consciousness. Yet if one listens closely to Francis and teachers such as those in the Cranton and Carusetta study, another aspect of the self is suggested: an intuitive, extrarational dimension of the self. These teachers seemed to be referring to something other than just the use of analytical, rational, and judgmental processes to think about their teaching. When they talk about reflecting on their teaching, they refer to feelings, hunches, intuition, and insights from practice. In Jungian language, these are expressions of soul. Hollis (2005) describes the soul as "our intuited sense of a presence that is other than the ego . . . the archetype of meaning and the agent of organic wholeness" (p. 253). It represents "an active place of wisdom, deeper than my conscious knowing" (p. 254). Soul represents the aspect of the self that animates our inner and outer worlds, bringing them to life (Hillman, 1975; Hollis, 2005).

As Francis walks the deserted hallway of the classroom building, his *thinking* and *reflecting* suggest activities of the self arising from an aspect of his being other than the analytical, rational, and critically reflective processes of ego consciousness. Even more important than the series of questions that pepper his consciousness, Francis seems caught up in waves of powerful feelings, fantasies, and emotion-laden images.

The emotional dynamics of teaching can often swamp ego consciousness and its critically reflective processes, as they seem to have overwhelmed Francis. Rather, it is more appropriate to understand his experience as how the soul expresses itself within the everydayness of our lives, requiring something other than critical reflection and analysis. Critical reflection can help us develop a greater sense of self, but we have to be careful around matters of the soul, and the experience of powerful emotions indicates we are in the presence of soul. Self-understanding is grounded in a deep sense of soul, of nurturing and caring for soul (Moore, 1992; Palmer, 2004). Palmer (2004) suggests that the use of critical, analytical methods often drives the soul into hiding.

Arising from and mediated by the unconscious, the language of the soul has the potential to tell us a great deal about ourselves as teachers and about our relationships with learners, our content, and the contexts in which we teach (Dirkx, 1997). As in the case of Francis, this "language" is mediated not by the rational self but by one's imagination; it is constituted by feelings, fantasies, and emotion-laden images.

Emotions, Imagination, and Authenticity

We foster authenticity in our teaching by connecting with a deeper sense of who we are. Doing this represents a kind of soul work, in which we attend to and work imaginatively with the emotions and feelings associated with our teaching, and the images that come to animate these emotions and feelings. Though they often remain unexpressed or invisible, at times powerful emotions and feelings erupt unexpectedly into the learning environment. Teachers are often caught by surprise by an emotional response to what is occurring within their relationships and interactions with learners (Dirkx, 1997). As with Francis, teachers might interpret student comments as a personal attack on their competence or character, creating feelings of anger, resentment, or annoyance. Frustration with a particular aspect of one's teaching can lead to the experience of anger and resentment in relationships with students, even manifesting itself in a general dislike of one's work and loss of heart (Palmer, 1998).

Although not a common focus of research in higher and adult education, feelings and emotions permeate the teacher-learner relationship, infusing it with meaning and significance beyond the rational and intellectual frameworks in which it is often studied (Dirkx, 2001; Robinson, 1996). We often think of emotions as potentially disruptive, but the experience of emotionality within one's teaching reflects what is important to one's sense of self and, ultimately, to development of authenticity in teaching. Through the experience of emotion, teachers come to recognize what is cognitively and affectively of value to them, and who and what they are. The experience of emotionality within one's teaching represents a language through which we can develop a deeper understanding of who we are as teachers and what teaching means to our life.

Our experience of emotions and feelings, however, are mediated through the imagination, and animated by particular images or fantasy figures evoked through our experiences. The imagination provides us with what Jung felt to be the only form of being we can experience directly (Davis, 2003). The human psyche is made up of images and fantasies that are evoked or elicited by particular stimuli or signs in the outer world. Often laden with emotions, these images represent a language of the inner self, through which we derive personal as well as archetypal meanings of the self and being in the world (Moore, 1992). For Hillman (1975), "Everything we know and feel and every statement we make are all fantasy-based; that is, they derive from psychic images" (p. xi). Through the agencies of symbol and metaphor, the imagination impregnates and illuminates the outer world with the images that populate the psyche. This focus on the role of the imagination in meaning making reflects an emphasis on the *mythopoetic* function of the unconscious (Watkins, 2000). That is, it stresses a capacity for myth making and generating and weaving fantasies that we then use to make sense of and guide our lives, often at an implicit and even unconscious level.

Our creative, active imagination offers us, if we choose to see them and work with them, spiritual guideposts to our own growth, healing, transformation, and development of self-knowledge. In soul work, development of self-knowledge and authenticity involves a conscious, imaginative engagement of the unconscious dimensions of the self. Hollis (2004) suggests that "we are imaginal creatures; through images the world is embodied for us, and we can in turn embody the world and make it conscious" (p. 31). The imaginative functions of the psyche serve to make visible unconscious, invisible energies within the psyche, thereby connecting the conscious self with what is more authentic within oneself. The imagination spontaneously creates the images that we use to make sense of our encounters in the world.

Developing authenticity in teaching involves attending to and working with these images (Hillman, 1975; Dirkx, 1997). Being fully present to them helps one connect with what is deeply meaningful within the everyday immediacy of one's life and with its broader story and movement—the mythopoetic function of the psyche. We "come to recognize that we are often bound to lifelong scenarios which silently but constantly reveal themselves through the conduct of our lives" (Hollis, 2004, p. 10). These scenarios reveal one's personal myth, "an individual creation that crystallizes from a subjective experience of impersonal psyche and becomes a way of life" (Bond, 1993, p. 10). These personal myths reflect an imagined, emotion-laden story that conveys deeply held unconscious assumptions derived very early in one's life. These stories and their associated images are created and shaped over the years within patterns that are deeper and more impersonal or that transcend the individual person. They come to represent powerful structures of psychic energy, which exerts itself on one's consciousness, sometimes creating feelings of being caught or swept up in powerful forces beyond one's control. Jung argued that each of us is intimately immersed in a journey of self-discovery (Davis, 2003).

Examples of such foci of energy within our psyche that may be evoked within a teaching and learning setting are issues of trust, authority, power, competence, intimacy, and self-concept. These issues illustrate "structures" of the outer experience of the educational setting through which we live and realize our individual lives as teachers and learners. They are outlets for expression of powerful emotional dynamics within the individual and the learning group. These emotionalities are often associated with particular images or fantasies, such as seeing oneself at risk of being rejected by the teacher or the group, of being perceived as incompetent or powerless, or of losing one's sense of identity within the greater whole. Their expression within such images usually represents emotional issues related to the immediacy of the here-and-now as well as issues that have come to make up aspects of our personal myth and life story.

Such emotional experiences reflect, in a mythopoetic sense, challenges that reveal dimensions of our unconscious being. Self-discovery and the development of a genuine self—becoming more fully human—requires that these aspects of the unconscious be more fully integrated into consciousness. As we engage our work as teachers, our life stories take on certain patterns, such as wanting to be liked by our students, being perceived as a competent and caring teacher, and making a difference through our teaching. As we reflect on our teaching, we may see that these issues become focal points for the experience of powerful, emotional dynamics and relationships and interactions with students, such as those conveyed in Francis's story. To be sure, his experiences were saying something about his present situation, but his feelings seem also to echo from and bring forth a past that seemed ever present in his life as a teacher, feelings of never quite measuring up or making the grade.

At some level, many teachers can relate to Francis's powerful emotions. Our initial reaction to such experience reflects a tendency to pathologize such feelings and analyze what might be wrong with our current situation. Development of identity, integrity, and authenticity (Palmer, 2004) within teaching requires critical reflection and the work of the intellect. But these faculties alone are not enough. Developing a deep sense of authenticity also requires a symbolic approach to such experiences. A symbolic perspective directs us to the images and fantasy figures associated with these emotional experiences and the way in which our life journey might be seeking expression through these images and experiences. The experience and understanding of emotionality and its associated images becomes a medium through which we develop and foster self-knowledge and authenticity in teaching (Dirkx, 1997; Moore, 1992). This process requires us to attend to these various images, allow them to come to life, honor and animate them as aspects of oneself, and give them voice in our lives by engaging them through imaginative dialogue.

In the remainder of this chapter, I focus on how we might foster authenticity and a deeper sense of self by imaginatively engaging the emotion-laden experiences of teaching.

Fostering Authenticity Through Imaginative Engagement with Teaching

In her book on becoming an authentic teacher, Patricia Cranton (2001) highlights some of the struggles involved in trying to be a good teacher and grounds her discussion of developing authenticity in the importance of self-understanding. Cranton encourages us to delve "into the center of the Self" (p. 2). In this chapter, I lay out an approach to the self and self-understanding grounded in the concept of soul and soul work. This sense of authenticity reflects commitment to a form of professional development that engages the self in an imaginative, dialogical relationship with unconscious dynamics of the psyche, the work we do within the transitional space between experience and its deeper meaning (Peppers and Briskin, 2000).

The process advocated here for developing authenticity in one's teaching entails reflective work and is enhanced through some form of journaling process. Use of the journal is a powerful technique for the disciplined practice necessary to recognize and imaginatively connect with the various aspects of the self and help us make sense of our life experience and journey. Experience and the symbolic approach form the foundation for professional development that aims, through soul work, to foster authenticity in teaching. I begin with working with the concrete experiences within teaching and then fold in more symbolic approaches.

Use of Specific Experiences. Concrete and immediate teaching experiences represent our text for this approach to professional development. As Francis walked the hallway after class that evening, he demonstrated the early stages of engaging the experience; or perhaps it is more accurate to suggest that various aspects of his teaching experience that evening engaged him. Stories of the soul often reflect disappointment and a loss of confidence that can leave us feeling quite troubled about our career and our life (Palmer, 2004). As Peppers and Briskin (2000) suggest, "The journey of the soul often begins with the experience of being lost" (p. 22).

In working with our experience, it is important to suspend judgment; become active observers of our own actions, behaviors, emotions, and feelings; and refrain from framing our reaction to our experience in terms of good-bad or right-wrong. In focusing on particular emotionally laden experiences, such as those Francis reported, we want to first recapture the experience—to describe as fully and completely as possible what happened and what specific emotions and feelings were evoked. We want to be sure to account for all the key players involved and how they contributed to the significance of the experience. We want to remember as best we can how we felt about each person, the emotions we experienced with his or her actions and behaviors (were we angry, hurt, feeling betrayed, disappointed, confused?). As we attempt to connect with the specific experience and recall (best done in writing) what actually took place, we also want to attend to the feelings that are engendered as we revisit the experience—feelings of defensiveness or willingness to be open, curiosity, and justification.

We should also not be afraid to let ourselves wander because the psyche may be taking us places the conscious ego would rather we not visit. Our goal here is to give voice to non-egoic dimensions of our being and rely on our intuitive, imaginative capacities, such as fantasy or daydreaming. It is helpful, though, when we become aware that we are engaged in the fantasy, to take note of the fact and gently return to the generative image or experience with which we started.

In reflecting on his experience of that evening, Francis has already begun to recognize how this particular interaction seemed to draw his attention to themes in his life that stretch back across time. He has acknowledged some of the "impersonal" dimensions of his experience that were derived from events much earlier in life but that continue to shape and give meaning to concrete situations of his current life, especially his teaching practice and interaction with certain students.

In doing so, Francis has also taken another important step with respect to working with the disorienting experience: locating the experience within the broader context of his own personal myths. Similarly, as this occurs we can ask ourselves of a particular experience, Given what we have learned about ourselves as teachers and the broader movements of our lives, what aspects of that experience would we be willing to give up? What do we trade, and why?

Use of a Symbolic Approach. Another approach to establishing and working with a dialogical relationship with the unconscious is to make conscious use of a metaphorical or symbolic approach (Peppers and Briskin, 2000). By this I mean using language such that it helps us to perceive an aspect of our being that may not be displayed in full view through our experiences in the concrete world. Use of a symbolic approach helps us begin to glimpse aspects of our being that are hidden from conscious view. The symbolic approach affords this level of perception through metaphors we use to describe our own experience; metaphors used by others within our experience but seeming to evoke strong feelings in us; or metaphors or images that may be embedded within books, stories, or articles that we read. In using this approach, we take note of our reaction to particular metaphors, symbols, or images—what our attention is drawn to—and our emotional reaction to these images.

In reflecting on our experiences, we might at times feel like the high school kid in gym class chosen last for volleyball. Although Francis does not explicitly offer a specific image arising from his experience, when asked he might volunteer that the experience made him feel like an orphan desperately wanting to be loved and accepted but again feeling rejected. Sometimes, when I feel pressed by simultaneous demands from multiple students, I often fantasize about getting on my motorcycle and, with a small pack, heading off into the sunset, leaving behind all my messy cares and worries, with the wind and the sun as my companions. As we take note of and record these images, we want to attend to what they say about ourselves, what feelings about ourselves or other people are suggested, or how the images relate to aspects of the role we play in our work as a teacher.

In addition to attending to the images evoked within my experiences of teaching, I also read life stories, particularly ones reflecting individuals struggling with the existential questions of life, work, and love. Such stories constitute a rich library of evocative images that pull me more deeply into the lives of those about whom I am reading, while at the same time pulling me more deeply into my life. When read with a journal nearby, such works help foster a dialogical relationship with the unseen in life. In reading these works, we can approach the written text similarly to how we would approach the "texts" of our concrete experiences.

Recently, I completed a lengthy text by Paul Elie (2003) on the lives of Dorothy Day, Flannery O'Conner, Walker Percy, and Thomas Merton, titled *The Life You Save May Be Your Own*. The story weaved the lives of these four individuals and their struggles with questions of identity, faith, and work. I am drawn to biographies and autobiographies, such as *A Life in School* by Jane Tompkins, that focus on self-reflection as a medium for deeper learning and self-understanding. Works of fiction by contemporary authors such as Margaret Atwood, Richard Ford, and John Updike, or classical authors such as Dostoyevsky or Tolstoy, are a similar medium. Certain works of poetry or poetic prose, such as the work of Annie Dillard, also present images that help mediate access to aspects of my self not readily apparent to waking consciousness. For similar reasons, I use such resources not only for my professional development but also as texts in my teaching.

Use of Fantasy. Finally, we can foster authenticity within our teaching through fantasy (Peppers and Briskin, 2000). Fantasy and animation allow us to play with images as characters, opening us to more aspects of our unconscious selves seeking a voice in our lives and our processes of self-knowledge. In making use of fantasy in our professional development, we make explicit use of our imagination to create or develop stories that reflect various aspects of our psychic life. We can use fantasy with the emotions and images evoked by our concrete experiences or metaphorically through images we encounter in the texts we read.

For example, if we have carefully described a disorienting experience, as outlined earlier, we can continue to work with one or more of the images identified in this experience through fantasy. Francis might have gone on to describe the orphaned-child image. In a fantasy exercise, he might give the orphan a name and imagine his life as a grown adult. In other words, Francis would create a work of fiction by free-associating this image and writing about what comes to mind. What was the orphan like as an adult? What did he look like? What was his life work? What issues did he struggle with in love and work? He would do this within a set period of time—say, ten or fifteen minutes—writing nonstop during this period but then stopping at the allotted limit.

Writing the story is the first, important step in use of fantasy. Similar to the text of experience or the text of fiction, however, we need to play the role of observer of our own story. We reread the story and make a list of questions about it. What parts of the story make us uncomfortable? What aspects bring us joy or delight? What surprised us about the story? In reflecting on the story, we seek to engage both thought and feelings. We want to know what the image or character felt, what emotions are experienced as we reflect on the characters and their stories.

Finally, in reflecting on the story, we want to take note of what about it is reflected back to us. The material that is evoked through this process represents what Peppers and Briskin (2000) refer to as "our own well of images" (p. 38). This process helps to establish a conscious, dialogical relationship with these images, to develop deeper awareness of our emotional life, and to potentially foster authenticity related to our teaching.

In addition to these personal and self-directed approaches to working with images and the self, more formal professional development opportunities are available that permit a more social and community context for this work. Examples are Parker Palmer's Courage to Teach Program (Palmer, 1998) and Ira Progoff's Intensive Journal Workshop process (Progoff, 1992). These programs are structured, guided, in-depth experiences with soul work. They represent an environment that respects the deeply personal and individual nature of this work yet makes use of the powerful dynamic of the community of learners to help foster one's development.

Conclusion

The teaching-learning setting represents a social context that requires a symbolic perspective as well as a literal one. The quest for authenticity, however, is ultimately symbolic. Often shrouded in thick mists of uncertainty and ambiguity, its path is strewn with images, symbols, fantasies, metaphors, and dreams, the living and the dead, the past and the future hopelessly intertwined with the present. In the symbolic approach, "what is essential is invisible to the eye" (Saint Exupéry, 1943, p. 63), the development of authenticity rests with our willingness to muck around in the dark, messy, unpredictable world of the unconscious.

I have shared the story of Francis in several contexts, and participants readily identify with various aspects of his experience. We all seek meaningful work that makes a difference. Many of us believe that teaching represents the potential to realize such desires. Yet, as so many studies and stories suggest, teachers lose heart and end up teaching to the test rather than from the heart. The craft of teaching is intimately bound up with who we are as a person. We can teach from the heart only if we recognize its role in our life and develop a conscious relationship with it. As much as many policymakers and politicians would like, we cannot separate the art of teaching from its artisan. As a goal for professional development, developing authenticity in one's teaching is barely a blip on the radar screen. Yet on the horizon one can discern promising and hopeful indications that this state of affairs is changing, and that the inner work of the teacher is at least as important as the outer work of developing texts and managing classrooms.

New Directions for Adult and Continuing Education • DOI: 10.1002/ace

Cranton (2001), Palmer (2004), and others who encourage us to ground our teaching practice within a deep sense of identity, integrity, and authenticity are asking us to make a profound commitment of time and intellectual and emotional energy to this work. The journey to authenticity in teaching, like individuation itself, is a demanding trek not to be entered into or undertaken lightly. The ideas I share here represent powerful strategies for helping us connect with the inner self and expressing this self through our teaching. We should approach this task with a deep sense of respect, wonder, humility, and love.

References

Balint, M. *The Doctor, His Patient, and the Illness.* New York: International Universities Press, 1964.

Bond, D. S. *Living Myth: Personal Meaning as a Way of Life.* Boston: Shambhala, 1993.

Brookfield, S. D. *Becoming a Critically Reflective Teacher.* San Francisco: Jossey-Bass, 1995.

Clark, M. C., and Dirkx, J. M. "Moving Beyond a Unitary Self: A Reflective Dialogue." In A. L. Wilson and E. R. Hayes (eds.), *Handbook of Adult and Continuing Education: New Edition.* San Francisco: Jossey-Bass, 1999.

Cranton, P. *Becoming an Authentic Teacher in Higher Education.* Malabar, Fla.: Krieger, 2001.

Cranton, P., and Carusetta, E. "Perspectives on Authenticity in Teaching." *Adult Education Quarterly,* 2004, 55(1), 5–22.

Davis, R. H. *Jung, Freud, and Hillman: Three Depth Psychologies in Context.* Westport, Conn.: Praeger, 2003.

Dirkx, J. M. "Nurturing Soul in Adult Learning." In P. Cranton (ed.), *Transformative Learning in Action: Insights from Practice.* New Directions for Adult and Continuing Education, no. 74. San Francisco: Jossey-Bass, 1997.

Dirkx, J. M. "The Power of Feelings: Emotion, Imagination, and the Construction of Meaning in Adult Learning." In S. B. Merriam (ed.), *The New Update on Adult Learning Theory.* New Directions for Adult and Continuing Education, no. 89. San Francisco: Jossey-Bass, 2001.

Elie, P. *The Life You Save May Be Your Own: An American Pilgrimage.* New York: Farrar, Straus and Giroux, 2003.

Hillman, J. *Re-Visioning Psychology.* New York: Harper Colophon, 1975.

Hollis, J. *Mythologems: Incarnations of the Invisible World.* Toronto, Ont.: Inner City Books, 2004.

Hollis, J. *Finding Meaning in the Second Half of Life: How to Finally Really Grow Up.* New York: Gotham, 2005.

Mezirow, J. "Learning to Think Like an Adult: Core Concepts of Transformation Theory." In J. Mezirow and Associates, *Learning as Transformation: Critical Perspectives on a Theory in Progress.* San Francisco: Jossey-Bass, 2000.

Mizrahi, T. *Getting Rid of Patients.* New Brunswick, N.J.: Rutgers University Press, 1986.

Moore, T. *Care of the Soul: A Guide for Cultivating Depth and Sacredness in Everyday Life.* New York: HarperCollins, 1992.

Palmer, P. *The Courage to Teach: Exploring the Inner Landscape of a Teacher's Life.* San Francisco: Jossey-Bass, 1998.

Palmer, P. *A Hidden Wholeness: The Journey Toward an Undivided Life.* San Francisco: Jossey-Bass, 2004.

Peppers, C. L., and Briskin, A. *Bringing Your Soul to Work: An Everyday Practice*. San Francisco: Jossey-Bass, 2000.

Progoff, I. *At a Journal Workshop: Writing to Access the Power of the Unconscious and Evoke Creative Ability*. Los Angeles: Tarcher, 1992.

Robinson, D. D. "Facilitating Transformative Learning: Attending to the Dynamics of the Educational Helping Relationship." *Adult Education Quarterly*, 1996, 47(1), 41–43.

Saint Exupéry, A. de. *The Little Prince*. New York: Harcourt, 1943.

Watkins, M. *Invisible Guests: The Development of Imaginal Dialogues*. Woodstock, Conn.: 2000.

West, L. *Doctors on the Edge: General Practitioners, Health and Learning in the Inner City*. London: Free Association Books, 2001.

JOHN M. DIRKX is professor of higher, adult, and lifelong education at Michigan State University.

*Although authenticity makes one vulnerable, the author
believes that its impact on learning and on enjoyment of
the teaching and learning process justifies the risk. This
chapter describes how to offer a relationship to each student, focusing on appropriate caring for individuals and
for learning.*

Authenticity and Relationships with Students

Katherine A. Frego

I believe that learning is strongly influenced by a student's personal needs
and motivators; I know that needs and motivators vary among students, and
change over time. This leads me to reflect on how the relationship between
teacher and student may have an impact on learning. How is authenticity,
as expressed in my relationship to my students, related to effective teaching
and enhanced learning, which is my first priority as an educator?

I have not formally studied the scholarly literature in adult education;
therefore this chapter represents my own reflections and experiences, based
on twenty-eight years of teaching at six universities and viewed through a
much shorter acquaintance with the concept of authenticity. In this chapter, I articulate how my teaching practices are outcomes of my beliefs con-
cerning the nature and role of authenticity, especially in my relationships
with students.

Positioning Myself: Who Am I?

My pathway to tenured professor, with a regular teaching load that includes
courses from first year to graduate studies, was circuitous. For twelve years,
I alternated between short-term positions as research assistant and sessional
instructor. I stepped into various courses or parts thereof, usually far outside my specialty and often at short notice. Fortunately for me, I have never
been forced to choose between my dual loves: research into plant ecology
and teaching. They are two sides of the same coin: my own love of learning

NEW DIRECTIONS FOR ADULT AND CONTINUING EDUCATION, no. 111, Fall 2006 © 2006 Wiley Periodicals, Inc.
Published online in Wiley InterScience (www.interscience.wiley.com) • DOI: 10.1002/ace.226

about plants is not only what I use to fuel my teaching but also part of what I hope to invoke in my students. I feel successful when students comment, "I thought plants were boring, but this course was interesting because you made learning about them fun."

I see the processes of teaching and learning through the lens of my discipline (plant ecology): true learning is growth, and it is strongly influenced by its environment. Transformational learning is enhanced by a positive, constructive relationship between student and professor. Perhaps selfishly, my enjoyment of the process is vastly enhanced by such relationships.

Role of Relationships in Teaching

Like many students, I thought lectures and textbooks were redundant. As a professor, I recognize that a human educator differs from alternative sources of information (books, Websites, news media) in offering more than content. A human educator models processes that create the culture of a discipline: how information is gathered, constructed, and wielded, and how a member of human society integrates a range of values and behaviors into that discipline. Furthermore, the learner can interact with the human educator, and therein is a relationship—however transient, shallow, or unacknowledged. I believe that humans are social beings hard-wired to relate to others; we may even imagine relationships of types other than what exist. It therefore seems reasonable to explicitly offer each student the option of a constructive and authentic relationship with me, not as the ultimate authority on my discipline but as a committed, enthusiastic practitioner who personally welcomes and guides the student into the discipline's community.

If teaching is one part of the relationship between two or more adults, the second is learning. In this context, the value of authenticity, in the sense of being honest or genuine, seems self-evident: no relationship can be positive and productive if the participants are not genuine, or if the intent of one is to manipulate or deceive the other. Indeed, a relationship based on reciprocal honesty and good intention can create a synergy that far surpasses any alternative I have experienced.

The first criterion of effective teaching must be to care that learning happens but is coupled with caring about the learner as a person. This caring must be motivated as much by recognition of the intrinsic value of the learner as by pragmatic recognition that the learner's social and emotional state of mind influences learning.

Whatever the motivation for caring (recognizing that some ways of caring are inappropriate in an educational environment), I believe that the critical criterion of authenticity is that the learner perceives the caring of the educator as authentic. At the same time, I would expect learning to be impaired if caring is perceived as inappropriate or unwelcome; I expect also

New Directions for Adult and Continuing Education • DOI: 10.1002/ace

that this varies with the level of caring expressed, student maturity, and the genders and cultures of student and educator.

My personal priority (in keeping with my life view) is to strive to express appropriate caring for students as fellow human beings with individual strengths and challenges. Although my experience has been that most adult learners want to "be known," I must respect those who prefer to be anonymous or private; however, all students must perceive that I care for them, even though their response to my caring is their choice.

My professional priority is to express my caring for student learning, because I believe it can be a powerful motivator for learning. One student told me, "I didn't really care about how I did in this course, but I knew you did. I didn't want to disappoint you."

Ideally, the relationship is triggered by mutual goals; a student wants to learn with a mentor who is willing to share. In reality (and in contrast to graduate studies), most undergraduates do not choose a course on the basis of their respect for the professor, and we certainly have little say over who we teach. The relationship may begin somewhat unwillingly: the student must endure a temporary relationship in which someone he or she does not know (never mind respect) imposes apparently arbitrary standards that will severely inconvenience his or her life. The professor must also endure the forced relationship, sometimes succumbing to the feeling of casting pearls of wisdom into an unappreciative and even resistant bottomless pit. (Perhaps I am overstating the initial negativity of the relationship, but as a lonely terrestrial botanist in a department that specializes in marine animals, teaching a required botany course has influenced my perspective.)

The immediate challenge is trust. Who will make the first move? Few learners, especially young adults early in the process of individuation, are open and trusting before personal safety is established. Therefore, my lowest priority is self-protection. The default attitude of "I care for my students' well-being" makes me vulnerable to injury; however, one cannot be open to students when things are going well and closed when they are not. I am also mindful of the power disparity: a student can easily (usually inadvertently) hurt me to some degree, but a blunder on my part could injure the student far more seriously.

In my experience, a large number of young adult learners lack confidence; many are either afraid of failure or lack the motivation to try hard. Even if I care only about learning, it is my task as an educator to offer tools and encouragement to overcome obstacles and eventually assist the learner to do so independently as an active, self-motivated learner. Effective education is learning that transforms the young adult in two steps: first, by establishing both the safety and the boundaries that allow and encourage the student to become actively engaged, with minimal fear; and second, by developing confidence in the ability to learn effectively (and self-assess) so as to create an internal personal safety that allows engagement in new learning situations.

New Directions for Adult and Continuing Education • DOI: 10.1002/ace

Scenarios, Anecdotes, and Examples

In my teaching, I strive for integrity in the literal sense of the word: to integrate my personal and professional life values, modeling them in and outside the classroom. As a result, there is no discrete boundary between my teaching and the rest of my life. Although I am not able to define my values explicitly, terms such as "honesty," "gentleness," "respect," and "nurturing" are good approximations. My contribution to the world is through my interactions with individuals. My caring for students is, I believe, genuine and part of my life view, but it is couched in words and actions that relate to their learning—explicitly within the discipline, unless they indicate that a broader interest in their overall growth would be welcomed, and even then with cautious respect for their (and my) right to privacy. Maintaining a balance between nurturing care and respect for privacy requires constant vigilance and creates a distinct tension in my life as I struggle to balance energy and time for my students with the same for myself and other people in my life.

Here is a list of strategies ordered by my sense of "appropriateness"— that is, there are some expressions of caring well within the bounds of standard pedagogical values, while others require more caution or present more risk. For example, in my mind it is impossible to be too clear or too honest (although Stephen Brookfield argues otherwise in Chapter Two), but one can be too nurturing (I admit to difficulty in maintaining balance). These strategies are described in the context of undergraduate teaching; however, I consciously apply the same principles to mentoring undergraduate summer field assistants and graduate students in research.

Attending to Student Needs. One universally appropriate expression of caring is to attend to students' perception of their academic needs. Throughout the term, I explicitly invite input on the course and in a variety of ways. I regularly hand out index cards and ask what I should start, stop, or continue doing; then I summarize the replies and my responses in the next class. I stop during lectures to ask, "Am I making this clear? Does this work for you?" I track the "pulse" of class, watching eyes and body language and adjusting the pace or approach appropriately. This may be a subtle shift in delivery, or a more direct "I see some puzzled looks. Shall we take that concept again from the top?" (I have been surprised to learn from students that they take note of my awareness of "classroom climate.")

Clear Expectations. A second appropriate expression of caring is to foster a fair learning environment, with clearly communicated targets and no hidden agenda. This entails explicit learning objectives for the course and for each component activity, both in the syllabus and orally as each item comes up, with a rationale—why it is needed, why this approach was chosen, its value in apprenticeship. The evaluation scheme is expressed as "How will you show me what you have learned?" I explain formative and summative feedback in terms of what each contributes, and how they can be used to increase learning. I state my philosophy about grades: learning is an individual's end product, so class mark distributions do not necessarily follow a bell curve. There is no reason a group of motivated students cannot all achieve A-level learning. (Sometimes I challenge them to do so well that I will be called to account for a "skewed mark distribution.") I make expectations as transparent as possible; for example, my grading scale is clarified quantitatively and qualitatively ("An *A* means that you are able to ...."). In each course syllabus and at the first class, I outline the Rules of Engagement as reciprocal rights and responsibilities of student and instructor. Specifically, they address fairness, effort, and respect. For specific assignments such as seminars, I engage the class in developing the evaluation rubric.

Valuing Individuals. In contrast to the two preceding expressions of authenticity, other strategies require sensitivity to students' rights to privacy and anonymity.

In terms of valuing individuals, many students want to be known—safely—and I must be the first to communicate "I care about you; you are important" without requiring, or even expecting, that they will reciprocate. Some simple behaviors contribute to this. Before each term, I send a global e-mail to the class list expressing my anticipation of working with them. I greet returning students by name and welcome all new students. I respond to e-mails immediately, with a personal tone. I invite office visits. I inquire about academic issues and follow up on topics previously raised by students. Questions that could threaten privacy or put the student on the spot are inappropriate; instead, I signal openness, safety, and welcome.

I also "open windows" through which the student may choose to be known. For example, in the first class I speak of the process of learning as a team effort, and I note that teams are more successful if they know each other. To that end, I invite students to introduce themselves. I myself begin, to set a nonthreatening tone, typically telling them about my household of pets. Most reply in a similar vein, often with good-natured laughter. Later, I may insert comical photos of my pets or family into my PowerPoint lecture. Insight into my home life often leads students to a sense that it is safe to tell me about issues in their own lives—for example, moving out on their own, conflicts with parents, and struggles with school work. This openness is intended to make students feel accepted and safe, but it requires absolute consistency on my part. For example, if a student misses a class to deal with personal matters, my reaction must be consistent both in terms of the values I have expressed in previous encounters and in terms of conveying personal safety—sincere empathy for emotional impact coupled with joint problem solving to address academic consequences.

One of my proactive solutions to the tension between real life and its academic consequences is "the amnesty coupon," included in the syllabus of upper year courses (Figure 4.1).

New Directions for Adult and Continuing Education • DOI: 10.1002/ace

Figure 4.1. Amnesty Coupon.

In recognition that we ALL have "real lives" and that sometimes a delay is unavoidable, I offer you a one-time amnesty from a deadline. You can redeem this coupon if you need to (be sure to read the fine print!), and you don't even need to explain. In fact, I hope you won't explain! However, if you manage to stay on top of things all term, you can redeem this coupon on the last day of class for a BONUS 2% on your grade.

This coupon good for extension on one assignment or lab report in BIOL 3541: Plant ecology. Offer void for tests & exams. To redeem, tear off, sign, & present to instructor. **No reason required.** One coupon per student.

Student:

Assignment:

Extended to (date):

Unclaimed coupons may be returned for **2% bonus** marks during the last week of term. Invalid where prohibited by law.

The coupon reduces the frequency of situations in which I must evaluate the legitimacy of an excuse and rewards those who are good time managers.

Caring. Caring about students requires clear professional boundaries. For example, I can acknowledge the impact of a student's personal problems and validate the student but then encourage, or even facilitate, qualified professional counseling or medical care. It requires sensitive antennae to detect the potential for an unhealthy or inappropriate level of personal revelation or dependence and to take constructive but evasive action. I also maintain close communication with campus counselors, so that I can signal the urgency of a request for an appointment.

Though I must not expect reciprocal caring from students, I can demand behavior that is mutually respectful of the learning neighborhood. This is outlined explicitly in my syllabi as well as in class. For example, I remind students that we have a contract: I will not be late for class and therefore disrespectful of their time, and they will reciprocate. I will not make disparaging personal comments about their behaviors or beliefs; nor will they, to their peers or me. I have noted increased responsiveness when such issues are expressed in this context of reciprocity. Perhaps it brings a revelation of what would happen if I, as a professor, treated students as they occasionally treat peers.

Reducing Anxiety. The most effective expression of caring has been my effort to reduce anxiety induced by fear of failure and power inequity. The relationship between instructor and student is rarely balanced—and probably should not be. However, a perception of lack of power may induce uncertainty, anxiety, and even outright fear, especially in junior learners. A moderate level of such stress can be motivating, but for many it reduces the

New Directions for Adult and Continuing Education • DOI: 10.1002/ace

willingness to take risks, and when extreme it can be paralyzing. My approach is to acknowledge fear as a reality, point out its impact, and furnish tools to overcome it (modeling their use). For example, in developing oral communication skills, I point out the ubiquity of nervousness in public speaking. I tell humorous anecdotes about my own learning curve, my failures (reenacting particularly disastrous scenarios), and my efforts to improve my "performance." I demonstrate the use of humor to defuse my own fears, laughing at myself.

Empowerment. Empowerment of students in the learning process can overcome anxiety and encourage autonomy and increased maturity in learning. Components of this strategy include building a student's confidence by promoting opportunities for success, expressing realistic assessment of the student's strengths and weaknesses, and encouraging "stretching." When students show potential in specific areas, I encourage them with new opportunities, or by facilitating contact with potential employers or projects. For example, students with artistic talent have created posters and Websites for local organizations, and I have hired many to assist with my own research projects.

Because many students see a chasm between themselves and those with expertise, I believe it is imperative to show them the bridge: experts are just expert learners. One powerful strategy is to encourage sharing of victories and frustrations. This can be used not only to develop the student's understanding of professional activities but also to enhance learning through acknowledging the role of personal life in academic and career pursuits. Furthermore, such sharing can extend beyond student-educator to student-student, contributing to development of a sense of neighborhood in a community of learners.

First, I use modeling. I share my own victories and frustrations in the context of professional activities relevant to my discipline. For example, I illustrate a class on scientific method with my own experiences, noting my personal reactions to parts of the process—the stress of grant writing or receiving feedback on a manuscript, and the elation of having something published. There is a simple academic value to this: students tell me that personal vignettes make the material real, and "reality sticks." Even so, students (especially young women) also report feeling encouraged when they learn that I am a professional scientist in spite of failures and setbacks, interpreting this as encouragement through empathy: "I have experienced similar challenges, and I know you can conquer them too."

A clear target is easier to hit. When students are asked to risk, as in learning a new skill, they frequently express anxiety about evaluation. If they can internalize the criteria by which learning will be evaluated, anxiety is remarkably reduced. I first applied this approach in required oral presentations in a second-year course, where the class develops a rubric for evaluation. I use humor to defuse initial anxiety in an "antimodel," a short presentation that is deliberately dreadful. As the class enthusiastically points out all its shortcomings, the criteria for ineffectiveness in a presentation emerge. It is a short task to convert this to a list of criteria for an effective one, after which we discuss the relative importance of the individual features. (For example, correct content is more important than attractive graphics; being audible is absolutely critical.) The class quickly reaches consensus on characteristics to include in an evaluation rubric, and percentage values for each one. Students frequently report they are surprised to learn that they recognize the characteristics of a good presentation, and they find it much easier to hit the mark if they themselves have drawn the target.

Nevertheless, self-assessment is rarely accurate in young adults, and overconfidence can be as dangerous as fear. In hopes of developing assessment skills with practice, all my courses require students to provide feedback on their own and others' performance. This is most explicitly developed in the oral presentation exercise described earlier. The evaluation is compiled from feedback by the instructor, by the student's peers, and by the student himself or herself. (Interestingly, classes always decide that peer evaluations should not count toward a final grade—although I have noted that peers almost always give higher grades than I do.) Each student's feedback session begins by asking the student to self-evaluate. What did you do right? What would you do differently? How might you work to develop this? The self-assessment is then calibrated against a summary of peer and instructor feedback, leading to explicit discussion of the student's individual strengths, weaknesses, and self-perception. Typically, students find that they already possess the basic skills needed and that they are highly critical of themselves in most areas but have "blind spots" and areas to improve. Our most productive conclusions are the student's own suggestions of what to do next. Students have reported that using feedback and assessment from self, peers, and instructor also opens their eyes to a variety of real-life issues, including types of feedback that are constructive (or not), ambiguity in or contradiction between evaluation criteria, discrepancies among evaluators, and ways to deal with criticism.

The oral presentation is also an example of encouraging students to stretch. Although many are terrified at the prospect, I work hard to reduce sources of anxiety (with attention to group size, evaluation criteria, and impact on grade). I also assure them that they are not evaluated on their level of confidence, and that no student has died (yet) from nervousness. By alternately pushing and pulling, as needed, every student in my required second-year course has presented and succeeded, and several have later reported that overcoming this particular fear opened new and exciting career possibilities.

Choice. In each course syllabus, I allow a measure of choice for evaluation. In my large second-year classes, students can retroactively shift the worth of midterm test versus final exam by 5 percent, to increase the value of the one on which they do better. Although my calculations show that it rarely changes a course grade, this relieves performance anxiety. In smaller upper-year courses, students select percentage values on individual learning contracts with a choice of required and optional items. We spend part of one class discussing expectations for each item, to assist in selecting a grade value that reflects effort and learning. For example, students may be required to do a laboratory project on a topic of their preference and then choose to communicate it to classmates as an oral presentation, poster, Web page, or written report. Because students select the value of these components, they may try out a new skill (such as Web design) with minimal risk to their final grade. Along with the choice, students also take more responsibility for deadlines and recordkeeping; I warn them that they should not expect me to keep track of all the contracts. Nor should they trust me to avoid mistakes in recordkeeping; they record their grades on their contracts and check them against mine at the end of the course. The variety of assignments is much less onerous to evaluate, and quite easy to administer; one spreadsheet contains the individual elements, marked on a consistent scale (for example, all presentations out of twenty). Individual students' percentages are calculated automatically on a separate but linked spreadsheet that contains their contract values (as low as 5 percent, up to 45 percent).

Offering choices extends to my tests and exams, to a varying extent. Again, students report reduced test anxiety in knowing that there will be a choice of questions. Through design of question type, this also allows me to test the same content areas in ways that match a range of learning styles, or to evaluate the use of a particular skill or approach across a range of topics.

Graduate Students

Because graduate students are typically older and often more mature learners, my approach to relationships with them entails a much greater degree of openness. Our graduate program is based on research, and my research area requires living in rustic conditions in remote forested areas for weeks at a time; it would be difficult to remain aloof. I apply all the general approaches described in this chapter, albeit in different formats; I describe one aspect of my supervisory philosophy as an example: the initial stages of building a relationship with a new graduate student.

Because I believe that an ideal graduate program should be a positive experience for both student and supervisor, I strive to work with students who value and expect what I have to offer, and vice versa. We begin with telephone and e-mail communication during which I point out that neither day-to-day research activities nor my style of supervising suits all graduate students. The candidate is instructed to exercise due diligence: I provide contact information for my current and past graduate students, encourage him or her to ask questions about what it is like to work in my lab, and assure him or her that all questions and answers will be kept in confidence. If this goes well, I invite the candidate to visit so that we can interview each other, using my standard list of values, priorities, and expectations to assess what I term "goodness of fit." I typically invite the candidate to stay with my family for several days during this interview process, and I host a potluck dinner for my graduate and honors students and lab staff. This allows the students (present and candidate) to form an opinion on whether it would be a comfortable fit.

This approach of building relationships within my lab group is an important element of the teaching and learning process. I do my best to maintain open communication with regularly scheduled meetings, one-on-one and as a group. I also work to maintain a sense of collegiality, encouraging students to help one another (and me) and share information and equipment. Aside from having a remarkably enjoyable history as a graduate supervisor, I have had few interpersonal conflicts with my students.

Conclusion

If I am authentic as a teacher, I know of no other way to behave. I have chosen to be authentic in my relationships with students in ways that I believe increase learning—by motivating, or engaging, or reducing anxiety and increasing confidence. In an interesting circularity, the concept of authenticity validates the fact that I base my teaching on what I value.

Over the years, I will be interested in assessing whether or not authenticity per se enhances learning—or simply allows me to enjoy teaching within my personal comfort zone, which enhances learning for students who occupy the same comfort zone while having little or no impact on those who do not.

KATHERINE A. FREGO is professor of biology at the University of New Brunswick at Saint John in New Brunswick, Canada.

New Directions for Adult and Continuing Education • DOI: 10.1002/ace

5

This chapter deals with the constraints that may inhibit authenticity, whether explicit and known to all or more insidiously embedded in our educational culture.

Institutional Constraints on Authenticity in Teaching

Russell Hunt

In Parker Palmer's *The Courage to Teach* (1998) there is a story that offers some profound insight into what we might mean when we speak of authenticity in teaching practice:

> For twenty years Professor X had tried to imitate his mentor's way of teaching and being, and it had been a disaster. He and his mentor were very different people, and X's attempt to clone his mentor's style had distorted his own identity and integrity. . . . Professor X's story gave me some insight into myself . . . early in my career, I too, had tried to emulate my mentor with non-stop lecturing, until I realized my students were even less enthralled by my cheap imitation than some of my classmates had been by the genuine original. I began to look for a way to teach that was more integral to my own nature, a way that would have as much integrity for me as my mentor's had for him—for the key to my mentor's power was the coherence between his method and himself [p. xx].

Palmer's point is not that lecturing is a bad idea—indeed, a fundamental assumption of his story is that it's possible that great teachers can use lecturing almost exclusively (in spite of the common argument among theorists of teaching that lecturing is always a poor choice). What is central here is that the teacher in question had somehow failed for twenty years to realize that the way in which he was trying to teach was incongruent with his

NEW DIRECTIONS FOR ADULT AND CONTINUING EDUCATION, no. 111, Fall 2006 © 2006 Wiley Periodicals, Inc.
Published online in Wiley InterScience (www.interscience.wiley.com) • DOI: 10.1002/ace.227

51

own habitual ways of thought and proceeding; it was inconsistent with what we might call his personality.

This is not to say that it was dishonest. What it was, one might argue, was inauthentic. This is not an obvious term here. Let me explain.

Defining Authenticity

There are, of course, challenges involved in agreeing on what we mean by "authenticity" (as Dr. Johnson remarked about the difficulty of defining another difficult term, *poetry*: "Why Sir, it is much easier to say what it is not. We all know what light is; but it is not easy to tell what it is").

What *authentic* means as people normally use it is elegantly phrased (amazingly enough) in the Merriam-Webster Collegiate Dictionary's examination of its synonyms:

> AUTHENTIC, GENUINE, VERITABLE, BONA FIDE mean being actually and exactly what the thing in question is said to be. AUTHENTIC implies being fully trustworthy as according with fact or actuality (authentic record).

Cranton and Carusetta (2004) have argued, however, that it is more than "the expression of the genuine self in the community"—that authenticity involves as well both critical self-awareness and "knowing and understanding the collective and carefully, critically determining how we are different from and the same as that collective" (p. 8).

I would suggest, on the basis of the extensive treatments of it by Cranton and Carusetta (2004) and others, that authenticity can usefully be thought of not as "honesty" but as coherence. This coherence or congruence needs to be internal, of course (does what I am doing match what I believe?). But it is also important that it be external (to what extent is what I am doing consistent with—that is, it acknowledges and respects but does not compromise with—the mores, structures, and constraints of the situation I'm in?). Cranton and Carusetta cite Taylor (1991), for example, as being highly critical of the modern ethic of authenticity, which contains the notion of self-determining freedom where individuals make judgments for themselves alone and without external imposition. Authenticity, like every other element of human behavior, does not occur outside of a social and institutional context.

It is not necessarily the case that authentic teaching is better teaching. It is entirely conceivable that a profoundly self-consistent and context-consistent teacher might be lacking in crucial skills; might not have sufficient understanding of her field; or in some other way fail to touch, influence, and shape students' understanding. Perhaps less obviously, it is also possible to imagine teachers who are carrying out what they see as the demands of the situation unreflectively, where their actions, as Jarvis phrases it, are "controlled by others and their performance is repetitive and ritualistic" (cited in Cranton and Carusetta, 2004, p. 8) but, in a given context, have a consistently powerful and positive effect on students' learning.

What authenticity in teaching does guarantee, however, is the potential for growth and change, and the possibility (indeed, the likelihood) of experiencing, confronting, and dealing with discomfort. Palmer's teacher was stuck—as the rest of the story makes clear—in a way of teaching that did not suit him and that he could not change until he recognized its lack of connection with his own internal priorities and preferences—indeed, not until he recognized that this kind of internal consistency was imaginable, desirable, and attainable. Palmer offers as the simple premise of his book that "good teaching cannot be reduced to technique; good teaching comes from the identity and integrity of the teacher." Thinking about this from the perspective afforded by the concept of authenticity allows us to see "identity and integrity" as an idea complicated in ways we do not often consider. What is particularly important is that such a perspective invites us to consider how integrity is afforded and constrained by the situations teachers find themselves in, most centrally by the institutional context. Although I work in and write about the context of higher education, I believe that the issues I discuss here are relevant across most adult education contexts.

Institutional Contexts

When I talk about teaching with colleagues at conferences or over coffee, a common refrain of the conversation regularly has to do with what makes the job difficult—what institutional constraints, for instance, faculty labor under. Such complaints are often dismissed by university administrators or representatives of the wider public. It is easy to say (because it is true) that postsecondary teachers in institutions like universities and colleges are far less constrained than their colleagues in schools and community colleges. Comparing the freedom enjoyed by, say, a full professor of eighteenth-century literature with the onerous obligations dumped on a fifth-grade teacher in any public school should, it seems, preclude any whining about constraints by the privileged denizens of the ivied halls.

It remains true, however, that we university teachers live and work in situations in which our choices are radically constrained, often in ways we are not even aware of. I used to dismiss as rationalization my colleagues' explanations that they thought some teaching idea was a great one, but that their circumstances made it impossible to try it. The dean would never stand for it, they said; the department chair would veto it, the students wouldn't go along with it. I argued, for example, that in fact no one was going to force a tenured full professor to give a final examination and count it toward some arbitrary percentage of a final mark, even though the university calendar mandated such a practice; or that lecturing less than the required one hundred fifty minutes a week was not going to bring the wrath of the dean

New Directions for Adult and Continuing Education • DOI: 10.1002/ace

down upon them—that indeed, as I discovered in a short sojourn in administration, the wrath of a dean is a singularly ineffectual tool.

I still believe this, but I have changed my mind about the power of institutional and situational constraints and become much more aware of the potential such constraints have—especially when they are not consciously recognized or explicitly stated—to shape our behavior and prevent us from acting consistently with our own most important priorities, and further to hide from us by habituation the extent to which our behavior has ceased to be consistent with our own fundamental convictions.

Structural Constraints

There are a multitude of ways in which the conditions teachers work in shape and constrain what they can do. Some are obvious: physical constraints such as timetables, classroom configurations, enrollments. Although such constraints are often taken as a given—whether a class meets three times a week for fifty minutes, twice for seventy-five, or once for one hundred fifty, for example, was not (at least when I began teaching in university contexts) usually seen as under anyone's control; nor was the configuration of the classroom. In some cases (traditionally, for example, composition courses or seminars), explicit control over enrollment is exerted, but most often class size is as a matter of waiting to see what happens and then coping. Similarly, physical classroom configurations have, in my experience, only recently been considered anything one had much control over. Whether the room had fixed seating, ranked auditorium rows, or (rarely) movable tables was a matter rather like the weather, except that fewer people talked about it.

Unlike teachers in schools, whose physical space is explicitly part of their teaching and who by default "have" classrooms they set up to support their own teaching priorities, university teachers commonly are assigned whatever is appropriate—usually determined by enrollment numbers and by an impersonal administrative process. In my first decades as a teacher, the physical space—the nature of the seating, for instance, or the layout of the room, not to mention the availability of chalk or white boards and bulletin boards—I worked in was simply a given, to be coped with like a blizzard or exulted in like a few weeks of balmy sun.

It is clear, however, once you begin to think about the consequences of such circumstances for teaching, that they have powerful consequences for the extent to which a teacher can work consistently with her own internal priorities, intentions, and values. A teacher whose personal priority requires extensive individual consultation with students obviously needs to find a way to deal differently with a class enrolling hundreds of students. One whose deepest need is to explain clearly with diagrams is likely to be radically hampered by a room with inadequate visual display capabilities. One who sees engaging students in group process around shared tasks or enterprises as central will need to deal with the constraints imposed by a fifty-minute class schedule (similarly, one who lectures will find a three-hour evening class a particularly difficult challenge). Obviously, in such cases a teacher forced to teach in a way that does not fit her priorities either has to rethink those priorities (this is not always a bad thing, incidentally) or simply cope—a consequence that can lead to long-term unconscious acceptance of inauthenticity as a condition of employment.

At least as obvious are other institutionally structured constraints, such as explicit rules about course conduct, mandated course descriptions, textbooks, and outlines. These are not simply difficulties to be worked around but can constitute serious challenges to authenticity. A university rule saying every class must meet for the full stipulated time, or take up topics in a certain order and for a given time, or work with the designated chapter of the textbook at a particular juncture, can of course be coped with somehow. You can, for example, address the issue explicitly in class. Such rules nonetheless retain the power, over time, to distance teachers from the values they hold central. To think of the conditions around such issues as ones that are not so much decided as preexisting, matters to be accepted as intrinsic conditions under which one works, is to move toward handing professional responsibility over to unidentified, distant administrators, and to push aside considerations that may be fundamental to one's beliefs about learning and teaching.

Other institutional limits are less explicit: the systems of general curricular statements and structures, and the "credits" and "marks" that implement them. Consider, for instance, a curricular structure placing a course as a prerequisite for another course or program—say, a second-year course in a scientific discipline that presumes students will have intimate knowledge of a particular theory or problem, one an individual instructor in the first year might think of as peripheral or see as an interruption in a learning process. This can force a teacher to choose between what she might think a poor learning strategy, involving introducing an idea before students are ready for it, and a better one—for example, taking time to build a foundation for the idea. Again, many such constraints are regularly dismissed as endemic to the entire teaching situation. Of course, universities have responsibilities to ensure that there are ethical and intellectual structures in place; mandated textbooks and course outlines can be seen as a protection for the university and as an aid to instructors, offering at best an irreplaceable set of guidelines to help shape teaching and at worst a crutch for underprepared teachers and a usable tool for defining inadequate teaching. That such issues are a fundamental part of the teaching situation does not mean they do not have potentially unfortunate consequences and implications for the teacher's authentic practice.

Related, but perhaps even less open to interrogation, are fundamental institutional issues such as the elaborate structures we have built around credits and marks. Of course, it is regularly argued that we have to have marks, that the system would break down if students had nothing to work for and we had no way of evaluating them (for them, for each other, and for the world at large). Our institution could not survive without a system of credits, so that we can track what students have taken, establish structures of requirements that everyone has to follow and that can be transferred from one context to another, and that are commonly understood.

For example, one might be forced to recognize that the institutional expectations of what constitutes an amount of work and learning appropriate to, say, a three-credit university course make it impossible for a teacher to avoid telling a brilliant student who has not completed an independent study that she will have to be satisfied with her own learning as a reward, because there is no room in the agreed system of credits for her work to be included (at my own institution, for example, there are three- and six-credit courses and no provision for anything else). One might hold the belief that a learner should be allowed to drop out of a course at any time without penalty (without, for instance, incurring a failing grade on her transcript, which would lower her grade point average) but in fact be working in the almost universal situation where the institution's structures require a passing or failing grade. To continue in such a situation for years is either to continue to be profoundly uncomfortable or to make the adjustment, get used to it, forget about it, and become a teacher working in conflict with one's own fundamental values without knowing it.

How does one achieve authenticity in a situation like this? It is easy to say that simply recognizing and acknowledging the inconsistency should be enough, but over time it is difficult to avoid developing calluses on the more sensitive parts of one's identity.

Policy Constraints

Other characteristics of the postsecondary teaching context that pose challenges for authenticity in teaching include many policies and practices intended to promote "good teaching": university guidelines on teaching, structures for evaluating candidates on the basis of teaching for employment, promotion, tenure, and awards. Like many of the other pressures on teachers I have mentioned, these cannot be characterized as negative; indeed, many universities are justly proud of the progress made in recent decades toward properly valuing teaching as part of a university's central mission. It is important to recognize, however, that such measures and policies can have unintended and damaging consequences for individual teachers. Structures for evaluating teaching, particularly mandated and summative teaching evaluation forms, can create a situation in which teachers are impelled to adopt strategies inconsistent with their own deepest values.

Teachers who wish to challenge deep-seated assumptions among their students find that pursuing such challenges can generate a good deal of discomfort among the students, which in turn can result in "low numbers" on the teaching forms (see Brookfield's discussion of this issue in Chapter One). Similarly, teaching practices that render students uncomfortable—a common example is group work, which students who are used to the agonistic, individualistic practices of most school contexts often object to—can pose a serious problem for a teacher threatened by use of ratings in her tenure portfolio or her application for a job elsewhere.

A more general way of stating one aspect of this problem, as Cranton and Carusetta (2004) have observed, is that the usual institutional guidelines used to evaluate and promote so-called good teaching "most often provide principles, guidelines, strategies, and best practices, without taking into consideration individual teachers' personalities, preferences, values, and ways of being in the world—the ways in which they are authentic. The assumption underlying this approach is that what works well for one teacher in one context works well in general, for all teachers in all contexts" (pp. 5–6).

Social Constraints

The explicit expectations of good teaching expressed institutionally often reflect what is perhaps the most powerful of all constraints under which teachers labor, and the one that has the most potential to extort behavior that is inauthentic. The social assumptions around the role of "professor," for instance, among colleagues, students, and the public, have a powerful influence on what you can do, even though you may not regularly attend to or be conscious of them. Expectations about what happens in class sessions are built into the language we use to talk about them—regularly, for example, teachers (and others) still do call them "lectures," even if what happens isn't lecturing at all; and people hired to conduct them are often still called "lecturers." As has been often noted, the conventional language around our work refers to a "teaching load" (though never to a "research load"), and professorial positions are often described exclusively in terms of hours of teaching required per week—by which is invariably meant contact hours, or lectures. Such language "frames," to use George Lakoff's useful term (1987), any discussion of teaching, or indeed any instance of practice, in terms that may invisibly exert pressure on a teacher to work in a fashion tending inevitably toward what we have to call inauthenticity. The tacit shaping of thought and speech facilitated by such discourse profoundly affects students' assumptions about and expectations of teachers. It is easy to say that an important part of a teacher's job is to confront, make explicit, analyze, and alter such assumptions and expectations. But we are very much like fish in a position of having to "deal with" water: keeping the presence of these elements in consciousness is, in practice, virtually impossible, not only for our students but for ourselves.

Every teacher, for example, is subject to the virtually universal expectation of students that her language will be evaluative. The classic example occurs in moderating class discussion, where the teacher's standard move to promote further thought about an issue ("Yes. Anyone else?") is interpreted by the student as actually meaning "No. Wrong." As an English teacher, I have been aware for many years of how marginal comments on student writing are read—regardless of my intention—as a rationale for evaluation. "I don't understand" is understood not as an invitation for explanation or a recounting of a reader's experience but as a brick in an incremental evaluative structure, equivalent to "bad." Of course, it is possible that a given teacher, over an extended period of time with a given student or class, might be able to alter this first and fundamental character of the relationship; the tacit presumption that the teacher's fundamental role is an evaluative one remains a "default mode" against which one must unremittingly struggle. Or that one gets used to.

The postsecondary student's expectation of "teacher" is, in general, that the teacher decides what is to be learned, has the correct information, knows the appropriate theory, takes the right approach—is the source of truth, at least in that context, and further, that the teacher's fundamental job is to make sure the student has the right information, the right theory, the right approach, by judging the student as right or wrong. Such a model of what teaching is, of course, does not match the model held by many and perhaps most teachers, but it is remarkably persistent and remains a pervasive source of mutual incomprehension. Such specifics arise from a widespread social norm having to do with what a "professor" is, of course. This is where, to be authentic, individual teachers often need to separate themselves from that socially defined collective.

Its very persistence is a powerful force militating against authenticity. As with all these persistent constraints, it is quite easy to ignore incomprehension and slip into comfortable, tacit acceptance of it as an unavoidable consequence of the situation we are all in—and at the same time quite difficult to maintain the contrary stance, to continue to assume, and talk and write as though you assumed, that the teacher's discourse about the student's work and thought is authentically dialogic.

What makes this such a challenge is precisely the persistent universality of this model of teacher-student relations. Every student, every class arrives with expectations of this kind. This is perfectly reasonable, because it is based on their consistent experience with educational institutions. Because of this, forging a more dialogic frame for classroom discourse is a task that not only is never complete but is begun anew with each student, each class, each semester.

In my own teaching, for example, I reject certain tacit institutional constraints (with profound consequences for how my courses are conducted):

- That students learn how to write—and think—by means of feedback on their written work from professors
- That a student's understanding of a field or concept is adequately mea-

sured by her ability to write coherently, conventionally, and at length and in an organized way about it

- That the most powerful motivation a student encounters is the urge (or, more likely, the necessity, the requirement) to please a professor and thus get a good mark

To deny these assumptions is, I am aware, radical, perhaps foolhardy, and conceivably silly. I know that they shape much instruction and are adhered to by many others with passion and conviction equal to mine. Every term, the universal adherence to these convictions among my students must be somehow addressed without usurping the actual curricular goals of each course. In other words, a course simply cannot be transformed unilaterally from one offered and advertised as an introduction to eighteen-century drama into an exploration of the nature of learning. Yet ignoring them is impossible: a course that does not employ academic essays and term papers and does not provide for professorial feedback on written work courts simply being dismissed as an aberration. Requiring and marking term papers would put me in a position describable only as profoundly inauthentic.

In one way or another, I would argue, every teacher is in this position to some extent, and the problem is always to find a way to negotiate between these conflicting demands so as to allow the teaching practice to remain coherent (that is, in the terms I stipulated at the outset, authentic). That the situation may never be "solved" is, I have argued, not a bad thing: a continuing commitment to authenticity, to an ongoing negotiation among theory, practice, and context, is the best possible way to ensure continuing growth and change.

A recent experience in my own teaching raised a number of these issues for me, and it may help in making clear how they can present themselves in particularly challenging, and potentially productive, ways. In presenting this episode, I should be clear that I am not suggesting the teaching methods and theoretical assumptions underlying them are appropriate to others. I think, though, that they bring this process of negotiation among conflicting demands into relief.

The Knot

Let me begin with some background. For a decade, I have been involved in team-teaching a first-year cross-disciplinary eighteen-credit-hour learning community. The three of us most centrally involved in this particular section of the program have called our section "Truth in Society," and as part of our offering first-year courses in English, religious studies, and sociology we have organized, each year, collaborative student investigations of particular historical episodes in which, as we define the criteria, people's fundamental beliefs were challenged or changed. The idea is that by conducting an investigation of the episode, coming to an understanding of the beliefs and assumptions at stake, and presenting this understanding to the rest of the class, small groups have an opportunity to learn many of the things we want university students to know and understand—about libraries, writing, scholarship and research, critical reading and thinking—as well as encountering how the three disciplines involved in the section might address the issues.

Having agreed on criteria and methods for making decisions about events, we give students the power to make their final choice about which events are worth studying. In the ten years of the program, events or episodes have ranged from the fatwah issued on Salman Rushdie by the Ayatollah Khomeini to the Salem witch trials, from the Scopes trial to the visions of Bernadette at Lourdes and to the Brixton riots. In recent years, one of the strategies for identifying such episodes has been to ask the students, in their reading of current articles in journals, to watch for references to appropriate historical episodes (to make research more practical and useful, we established that the event had to have occurred at least twenty years ago).

In the fall of 2004, a member of the class found an article called "Why the Vietnam War Still Matters," which appeared in a journal called *In These Times*. She reported on her reading on a public Website and among other things noted a reference to the author supporting John Kerry "for taking a stand and telling the truth in 1971 after coming home from Vietnam." Others read her report and wondered what exactly had happened in 1971 (remember, these students had not been born then). Through a process of discussion, Kerry's testimony in 1971 became a candidate for an investigation, and a group conducted what we call a "feasibility study" to determine whether the event was a good prospect for fuller investigation. The two students commissioned to assess the feasibility of the investigation reported with an unqualifiedly positive recommendation.

This recommendation carried the day, and the event became one of the four that groups were formed to investigate and report on, with the aim of helping the rest of us understand how, in this case, people's opinions had been challenged or changed. The hope of the three of us overseeing this process was, of course, that in the process of investigating the event there would be rich opportunities for the students to learn, with our tactful intervention, important things not only about the sources and tenacity of beliefs in 1971 but about how they themselves come to accept beliefs, and how their own beliefs might be changed.

One of the things the five-person group conducting the investigation were to do was draft what we called a "descriptive overview" of their event. Their overview, we were concerned to see, exhibited some slippage from the original focus (this is, of course, common in such investigations):

> In 1971 John Kerry, a spokesman for the VVAW (Vietnam Veterans Against War), stated that the United States was guilty of war crimes committed during the Vietnam war. He basically summed up the findings of the Winter Soldier Investigation, which proved that heinous war crimes were committed

against POW (soldiers and civilians). John O'Neill, who has been attacking Kerry since his speech in 1971, is now an active member of the swift boat veterans which is a group of Kerry's sea mates from the Vietnam War. This group was formed by George Bush during his campaign for the 2004 Presidential elections. We are planning on looking at both sides of this story and studying how people have come to believe what they do.

At this point in the course, students are working in independent groups, meeting with one of the three teachers when it seems appropriate and reporting to the rest of the class on their work at intervals. They are also posting the results of their individual research on Websites, which allows others to read them. As we watched and offered what advice we could, it became apparent as the group worked that the increasing reliance on material found on the Web was leading them to see primarily sources arising in the context of the current election campaign in the United States, and that the dominant voices in those sources were those of the so-called Swift Boat Veterans, a group organized primarily to raise questions about Kerry's war record.

Though where able we tried, as persistently as we could, to suggest that other sources and other strategies might be helpful, we believed that our primary goal was not to ensure that the students came to believe what we thought was "the truth" about the Swift Boat Veterans, but instead would have the freedom to conduct the investigation in the way they thought best. Our conviction led us to believe that to intervene aggressively would turn the investigation from one in which they were pursuing understanding into one in which they were trying to do what would best please the professors. In the event, although at various times each of us involved raised questions about why the students were accepting as reliable certain Websites or discourses, the Swift Boat Veterans were as effective with our students as they were, simultaneously, with many American voters.

The culmination of this process is that each group produces a bound book, which is read by everyone else in the class and then "launched" in a group presentation and discussion. The conclusion of the final report of this group was, almost predictably, that Kerry was unfit to be president of the United States. Our repeated attempts to help them be aware of how their own views were being challenged and changed—and especially my own attempts, as a teacher of rhetoric and persuasion, to help them see that the sources they were increasingly relying on were tilted in a certain direction—were unavailing. The five students were focused on building a coherent document and absorbing their sources into it.

As a teacher, this outcome challenged my convictions in a number of ways. What was most important to me as I watched the train rumbling down the tracks toward this conclusion was my own sense that it was simply wrong: my own reading led me to the conviction that the "Swift Boat Veterans" was an organization created for the express purpose of disseminating a distorted picture of Kerry's war record and had little to do with the

New Directions for Adult and Continuing Education • DOI: 10.1002/ace

testimony before Congress in 1971. But in our postmortem over the Christmas holiday, as we prepared for the second term of the course, the three teachers agreed that to have done more to steer the investigation would have been to violate our own convictions about the best way for students to learn: that the short-term outcome, a misconception about a historical event and a document buying into a propaganda campaign, was far less important than the long-term lessons about autonomy, independent learning, and public responsibility that the students had the opportunity to learn by being given their head.

I remained, however, troubled by questions that do not seem to me to have easy answers. How to intervene in such a process without becoming someone you do not want to be? How easy is it to become that person, point out the sources that were ignored and the voices that were silenced, direct the students toward the publications that questioned their conclusions and, at the same time, forget that it was not what you wanted in the long run? There is no easy way out of this kind of situation, perhaps—in fact, no difficult and challenging one either. It is possible to say, as I have said to myself, that one must remain conscious of the problem, to preserve the discomfort and avoid allowing awareness of the disjunction between ideal and real to fade, but there is no guarantee that this particular oyster's discomfort will produce a pearl.

More generally—and not at all incidentally—it is important to be clear that constraints are not a bad thing. We cannot be without them, and they help us shape our actions. As Robert Frost remarked, "You have freedom when you're easy in your harness." But you do need to be aware of the existence and nature of that harness, and to be conscious of how it enjoins tasks and limits flexibility—and how it consistently, regularly, and inexorably invites us to be conscious of our own need to retain our authenticity.

References

Cranton, P., and Carusetta, E. (2004). "Perspectives on Authenticity." *Adult Education Quarterly, 2004, 55*(1), 5–23.

Lakoff, G. *Women, Fire, and Dangerous Things.* Chicago: University of Chicago Press, 1987.

Palmer, P. J. *The Courage to Teach: Exploring the Inner Landscape of a Teacher's Life.* San Francisco: Jossey-Bass, 1998.

Taylor, C. *The Malaise of Modernity.* Concord, Ont.: House of Anansi Press, 1991.

RUSSELL HUNT *is professor of English at St. Thomas University in New Brunswick, Canada.*

New Directions for Adult and Continuing Education • DOI: 10.1002/ace

6

With an increasingly international student body in universities all over the world, there is growing contact between teachers and students from different cultures. This chapter brings cultural issues to a more conscious and explicit level so they may be examined in the light of teacher authenticity.

Cultural Dimensions of Authenticity in Teaching

Lin Lin

> The phenomenon of modern man has become wholly appearance;
> he is not visible in what he represents but rather concealed by it.
> —Nietzsche (1873/1997), *Untimely Meditations*

In this chapter, I weave some important considerations of the cultural dimensions of authenticity into an integrated approach to various aspects of authentic teaching in higher and adult education. I attempt to bring implicit, assumptive, and embedded cultural issues to a more conscious and explicit level. Rapidly evolving technologies that act as globalizing agents help to shrink the world, bring far to near, and create new challenges for today's teachers and learners. Compressed into smaller, nonhomogeneous groups, the changing prioritization of our values often points to the need for reflection and cross-examination on who we are as an authentic person and teacher, what teaching and learning mean, how we communicate with and understand each other, what makes our lives and work meaningful for ourselves and others, and why.

I first discuss some perceptions of authenticity in a number of cultures. I then examine the values that are behind perceptions of authenticity in teaching. I use the key concepts highlighted by Cranton and Carusetta (2004)—self, other, relationship, context, and living a critical life—as a framework for my discussion. In preparation for writing this

New Directions for Adult and Continuing Education, no. 111, Fall 2006 © 2006 Wiley Periodicals, Inc.
Published online in Wiley InterScience (www.interscience.wiley.com) • DOI: 10.1002/ace.228

chapter, I interviewed fifteen individuals from China who were studying as graduate students in U.S. universities. I quote from their comments throughout the chapter. Although discussing the dimensions of authenticity in various cultures, I am aware that I may be interpreted differently from what I intend and might promote further stereotyping of people from other cultures. I would like to emphasize the complexity of the issue rather than categorize fast-changing Chinese and American value systems.

Perceptions of Authenticity in Cultures and Contexts

A behavior that is perceived as authentic in one culture may not be perceived so in another. For instance, standing or speaking up for oneself is considered authentic in one culture, but it may be seen as egotistical or shameful in another. Holding back one's own thoughts to avoid temporary conflict or for the benefit of a community is considered gracious and altruistic in one culture, but cowardly or even deceitful in another.

The subtle clues of our intentions and actions are sometimes misinterpreted because the variety of customs, beliefs, and privacy concerns lead us to judge or misjudge the other's authenticity. For instance, it is common practice in China that when a patient is diagnosed with a terminal disease the doctor will discuss the death with the patient's family but not with the patient himself or herself. The rationale may be that the dying patient can live more happily (and longer) without knowing about his or her impending death, and that the focus of attention is on life rather than the possibility of death. In the United States, doctors do inform their patients of their situation. However, it is not that one culture values honesty and the other not. In the United States, people usually respond to "How are you?" with "I'm fine" (whether they are fine or not) because "How are you?" is merely a greeting rather than a question to be answered. In China, when people are asked a similar question, they may spend quite some time explaining why they are fine or not. Just responding with "I'm fine" may be interpreted as impolite or dishonest. We not only disclose or guard information on the basis of cultural norms but also according to what privacy means to us. In China, asking someone's salary is a common conversation topic; in the United States, people do not ask others how much they make unless they are close friends.

Our interpretation of authenticity is based on our values and cultural expectations. We have frames of reference or habits of mind (Mezirow, 1991) about authenticity. In other words, authenticity is a relative term. Do I share the values of my community? Or to what extent am I compliant, reflective, critical, or rebellious about those values? An individual can be a member of several communities simultaneously: a family, a workplace, a nation. The micro value systems for each community can differ.

Self, Other, and Relationship in Authentic Teaching

Teaching is a communicative act. To teach means that one is to contact, connect, and build a relationship with another person or group of people. Being aware of oneself and being aware of others are both important in a relationship, yet the emphasis on one over the other affects priorities in our value systems, which can dramatically change how we see and act in the world.

Awareness of Self and Others. Here are two quotes, from an interviewee who praised one professor as authentic and criticized another as inauthentic in a Chinese college setting:

> Martin, who taught my senior year in college, is a teacher I recall as an authentic teacher. He demonstrated a strong desire to learn about the students and their culture by courageously challenging the institutional rules and authority that set boundaries for foreigners not to cross. For example, he wanted to eat together the same food with students in their dining hall. It threatened the policy of face saving and "privacy" protection of the local authorities. . . . He eventually managed to eat with students but not for long. He also worked hard to introduce American literature and lives to us students—some of the books he gave us to read are even controversial in his own culture—it showed both his courage to work for the benefit of his students' learning by taking risks and also his belief of being an educator. . . .
>
> A contrasting example was another young professor teaching us drama in our postgraduate program. He came to classroom without teaching anything useful or substantial and turned students into remembering some terminologies that we did not understand for the final exam. He purchased and wore silk Chinese traditional clothes to make him look like he likes the culture he was teaching within and had a well-known affair with an undergraduate young girl in the university and had a lot of controversies with the university authority. Although he successfully put on a play he worked hard to create for the entire university to view, he was not an authentic teacher to his students or the department he worked for.

Several values undergird the appreciation for the first professor, who was seen as taking risks in order to know and teach his students; trusting his students' intelligence by introducing controversy even from within his own culture; and setting a good model by being an altruistic, caring, and trusting educator.

Similar values also contribute to the criticism of the second professor, who was seen as being interested in his own benefit but not caring about his students. He passed on information that was not understood. Having an affair with a student not only broke a rule but also upset members of a culture in which such behavior was considered immoral. Wearing silk Chinese

New Directions for Adult and Continuing Education • DOI: 10.1002/ace

clothes was perceived as being ostentatious rather than demonstrating an interest in getting to know the people.

Yet observing the situation as an outsider, one can also interpret the second professor's behavior as authentic from *his* perspective: he taught his subject area in the way he knew best, worked hard to do what he considered a contribution to the school by putting on a successful play, and showed his interest in the people by wearing traditional Chinese clothes. Although he might be consistent and authentic in his own beliefs and behaviors, he was judged harshly by his Chinese students as an inauthentic teacher. In a culture where altruism is highly regarded and a teacher is expected to set the example, the judgment can be instantaneous and harsh.

This raises a series of questions. Who judges authenticity? Who can? Was the second professor authentic if he saw himself as such but his students did not? It is obvious in the case of the second professor that his actions undermined his students' trust in him. When the learner does not understand the teacher, or when the connections between the subject matter and what is taught are unclear, the learner starts to question the teacher's credibility and motives. Given this, how can the teacher and the student build trust and achieve mutually shared understanding? How can both the teacher and the student take on the responsibility to help one another understand nuances of values, question seemingly authentic or unauthentic behaviors, and challenge actions and judgments on the basis of their own values? Inclusive change and transformation start with awareness of and consideration for the impact of the context, and they continue with sustained dialogue and interaction between the teacher and the student in the process of unpacking the issues and assumptions.

Traditional Chinese values emphasize humility and altruistic behavior. Two famous Chinese sayings come to mind: "a hero is silent about his glories" and "a good wine is known in all corners" (meaning that if it's a good deed, it will eventually be recognized). Children are taught to put the common good first and self second. Related to this issue, there is an emphasis on respecting other people (especially elders and teachers) and their opinions and knowledge. Chinese tradition since ancient times, including the three major ancient philosophies of Taoism, Buddhism, and Confucianism, all emphasize harmony and holism (Allen, Hu, and Bahr, 2005; Tzu, 2003). When a conflict arises, it is one's responsibility to listen to others' perspectives and modify one's own view to reach a balanced resolution to the conflict. There is strong emphasis on taking the middle path, assuming a position of balance and peace, while not dwelling on who is right or wrong, or who is better.

My experience in the United States sometimes presents an opposite expectation. I am reminded constantly that I should speak up for myself and tell people what I do; otherwise, no one will know what I do and thus the value of my work will go unappreciated. Presenting oneself and arguing for one's ideas and achievements are important skills to master in the American culture. If one does not do so, one is seen as less confident, competent, or capable (and somewhat disabled). This phenomenon is to a degree related to traditional American culture as a capitalistic society, in which prevail a spirit of presentation, commercialization, and the enduring pursuit of individual success. Crouch (2001) points out, "Ours is a technological era that often defines itself and achieves commercial success by continuing to do a better job at making the unreal seem true" (p. 3). To achieve recognition, wealth, and success, an individual is urged to constantly originate and promote ideas and products (Bellah, 1985).

As a result, values, which are deeply rooted in the culture and social construction of societies, set the foundation for differing points of view. The Chinese emphasis on moderation, balance, and humility can present a state that is hierarchical, authoritative, cowardly, and oppressive. Balance can bury and even twist an individual's originality, creativity, and authenticity. In contrast, the American emphasis on individualism, competition, and commercialization can also blur the line between reality and fabrication. Neither the Chinese nor the American way, if uncritically embraced, is authentic. The values and expectations imposed on individuals from both cultures shape people who conform to certain ways of being, to repress who they are or to present themselves differently from who they are. The authentic being is concealed in order to achieve a socially expected image or persona; consequently, the gap of understanding between people grows.

This gap is reflected in teaching and learning expectations. It is not uncommon to see Chinese students remaining quiet in the classroom. Sometimes such quietness is interpreted as lack of skills, or not having original, independent, or critical thoughts of one's own. A professor once mentioned to me a paradox he saw in many Asian students (especially female) in being responsive yet risk taking. He observed that on the one hand it is courageous for students to leave their home and study alone in another culture; on the other hand, it seems difficult for the same students to take risks and be responsible for their own learning. The students seem to agree easily (and responsively) to what a professor says but seldom express what they think themselves; nor do they make claims of what they want to do with their learning.

As an Asian female student, I shall reply from my own perspective as to why I seldom speak up in the classroom setting: (1) I'm used to listening rather than speaking, thanks to my past school experiences; (2) I'm more interested in hearing what the professor and colleagues have to say than expressing my own views; (3) I like to be thoughtful in what I say so that I don't waste others' time; (4) I want to make sure I really understand what I hear before responding or making judgments; (5) I tend to give the benefit of doubt to another's opinion, so I naturally hold back my critique; (6) I do not like to go to extremes and prefer to take a balanced approach to the issue at hand; and (7) I don't want to be confrontational with my colleagues unless it is absolutely necessary.

New Directions for Adult and Continuing Education • DOI: 10.1002/ace

It is easy to notice from this list that the issue is likely related to habits of mind or cultural norms. If a balanced approach is taken and more attention is paid to what others think, a person is nearly invisible and her voice is often inaudible.

With these different and dynamic frameworks, the questions remain: How does a teacher relate to a student authentically? How does a teacher see beyond his or her own frame of reference, and understand the issue underneath the silence of a student from another culture? How does a teacher recognize capability and talent in the quiet student, give the student similar challenges and opportunities, and encourage independent thinking and risk taking in learning?

Pursuing answers to these questions requires a continual effort to examine, understand, challenge, and reexamine the changing self, others, and contexts. Hansen (2001) recommended "tenacious humility" as a stance with which teachers (and students) might strive to approach their work. Making oneself a better person and teacher and promoting the same in one's students, Hansen says, is an ongoing journey: "Tenacious humility creates conditions for teacher learning, for a 'deeper knowledge' of the 'necessities' entailed in 'good practice'" (p. 172). Following this line of thinking, I discuss two important aspects of authenticity in teaching: the influence of context and living a critical life.

The Influence of Context on Authenticity in Teaching

Many factors—for instance, culture, class, gender, age, generation, background, profession, and race—shape a context. We judge (or misjudge) a person's authenticity on the basis of what we think is socially appropriate behavior. One interviewee offered two incidents to describe moments when she felt authentic and inauthentic while studying in an American university:

> [I feel a strong moment of authenticity] when I come to appreciate the fact that as an international student, I am creating a new self identify by staying longer in another culture, but in a marginalized way. A moment of authenticity is to be truthful to the fact of this marginality: yes, this is the way I live, being marginalized or not. I come to appreciate that my values and how I view the world are products of this unique marginality, the reality of my world that I cannot escape. . . .
>
> An occasion when I felt forced into unauthentic behavior is when I was caught in having to speak strongly for the traits of my native people or country in their defense when those traits were challenged by ignorance of their context and complicated reality. I was made to think it was unauthentic because on those occasions, I would tend to defend for the traits, both the good and bad parts—that is, a desire to put them in the best light, while I was aware this was not entirely true. . . .

It is not a simple coincidence that a sense of situated existence vividly came through these two incidents. Discussions of authenticity often appear in works associated with the existentialist philosophy of writers such as Kierkegaard, Heidegger, Sartre (Kaufmann, 1972), and Maxine Greene (1988). For these writers, the conscious self is seen as coming to terms with being in a material world and with encountering external forces and influences quite different from itself; authenticity is one way in which the self acts and changes in response to these pressures.

These descriptions illustrate the influence of one's contextual reality. For the person I interviewed, there were at least three contexts working simultaneously: (1) the context of her growing up in China, (2) the context of the American graduate school she attended, and (3) the dynamic context of herself in her marginalized yet transforming reality. She not only struggled internally and made difficult decisions between the conflicting values that she had become acquainted with but also fought against and compromised with external expectations, prejudices, and pressures put on her in the physical context.

With the increasingly international makeup of the student body in universities all over the world, there is more frequent contact between people of contrasting cultures and consequently more opportunities for understanding as well as misunderstanding of one another. A crisis emerges when one relocates from a culture that has traditionally been concerned with "moral considerations of the consequences of social interactions . . . and has extended greater respect to the 'common good' and/or 'group benefit'" (Gu and Zhu, 2000, p. 13) to a culture whose underlying premise is to think for and advance oneself (Bellah, 1985). The individual suddenly experiences an ontological breakdown that is based on personally assumed responsibility. A similar crisis exists for someone going in the opposite direction, as in the case of the drama professor who was criticized as being inauthentic and self-serving.

The student example is unique in that she is an international student and thus possibly encounters additional obstacles; yet the sense of powerlessness and feeling of being excluded or marginalized exist in all students (and teachers) at all levels. In the adult education literature, it is a tradition to focus on empowerment of students. However, a teacher needs to know the source of powerlessness in order to empower or help students find their power, if we believe that power is prevalent and in the hands of every individual (Foucault, 1980).

The feeling of being marginalized or alienated is not a place in which anyone wants to put himself or herself; but this is the first step in examining one's familiar framework in order to learn and grow. Lifton (1993) articulated this well when he wrote: "Individuals can be alienated from themselves only because there is something in them to alienate. That 'something' has to do with authenticity, with meanings and human associations that, over the course of a life, one experiences as genuine. The protean quest, however flawed, enhances that authenticity" (p. 232). The situation may appear to be hopeless and the individual feels powerless, but there are unique opportunities given to the person (as someone from outside the culture) to positively influence the thoughts and actions of others. Unlike the fish, which is "the last to know she is in the water," the outsider is able to see differences in values and actions that people from inside the culture may not be able to see.

In some professions, we are expected to carry with us a certain persona or professional code. The teacher as a professional is viewed as the "engineer" or "doctor" of the soul and is held to a higher standard than is the general public in many cultures. This high standard can entail having more knowledge, being more intelligent, being a better person, having the ability to inspire others, or showing tremendous courage. Good teachers are often depicted in the movies as heroes or heroines (Dalton, 1999). In many cases, these heroic teachers are misunderstood, act in opposition to a social norm, and are forced to leave the school where they teach (as in the films *Conrack*, 1974; *Dead Poets Society*, 1989; *Educating Rita*, 1983; and *Mona Lisa Smiles*, 2003). In other cases, these heroic teachers are finally recognized, accepted, and celebrated after struggling and finally saving the students and the school (*Mr. Holland's Opus*, 1996; *Lean on Me*, 1989; *Sister Act*, 1992).

Where does a teacher's personal authenticity fit in with these images, especially given that they are often in conflict with one another? Here, I discuss various aspects of living a critical life through a culture lens.

Living a Critical Life

We make decisions and judgments according to what we think is true and authentic. Interacting with others, we look for trends, search for patterns, and frequently check the motives of others, consciously or subconsciously. We may not necessarily look at our own underlying assumptions, many of which are part of our a priori knowledge or proprioception (Bohm, 1996). It is only through frequent dialogue that we discover our divergence and unearth each other's presumptions. A critical life assumes a certain level of critical thought. Critical thought usually means both looking at the underlying assumptions that are present in any argument and going back to the original or basic requirements of a thing and its intended purpose. Culture and language are so closely intertwined that it is nearly impossible to extricate the underlying assumptions. Even within the same community, one person's "inexpensive" is another person's "cheap," one person's "hybrid" is another's "aberration," one person's "willpower" is someone else's "shameful egocentrism." Being reflective of our own language is probably the most difficult but important task in a critical life.

Rousseau sees the problem of authenticity as that of the relationship between being and appearance (Heckle, 1991). An immediate paradox arises when we look at human beings as socially constructed individuals. An individual's authentic being and ways of being are shaped by and evolving with natural and environmental forces as he or she relates to and negotiates in the world.

Being an authentic teacher implies one is to communicate and relate to students, be aware of students' perceptions and perspectives, and be willing to be critical of and transform her own and the students' values and actions. This is a process of becoming a responsible member of a society or culture. The dilemma emerges when the process of becoming an autonomous individual does not join the process of becoming a moral and responsible person in a far-from-perfect society. To be a moral and free person, Rousseau chose to "withdraw into himself and became a friendless wanderer because he felt that he alone was just, authentic, truly alive; that history and society would betray him if he abandoned his solitude" (Heckle, 1991, p. 7). An individual struggles between maintaining an authentic identity and developing it through internal reflection and external interaction with the world.

If we constantly change, no one knows who we are or what we stand for. A person is appraised as an authentic person only if he or she is consistently perceived as being so. It is not that the person does not make any mistakes; on the contrary, the person is seen as real and thus fallible. One interviewee well articulated what she sees as the drive for the growth of one's authenticity: "Authenticity and personal growth are closely related. I believe that authenticity comes from people who purposefully make meaning from the lives we live, the work we do, and the people we meet in the reality of an unfair or not-at-all-perfect world. It comes with experience, especially the experience that has twists and turns that constantly push us for the question of the meaning and worth of existence. . . .

The complexity of the dynamics comes when we examine to what extent we still keep certain traits we are born with, and to what extent we are different as we grow, and why. Being authentic implies being simple and sophisticated, stable and resilient, independent and interdependent, moral and critical.

Conclusion

The purpose of this chapter is to help expose assumptions and perceptions we hold about authenticity in teaching across cultures. Culture and language hold the context for these assumptions, and they are often difficult to discover, especially if they have been seamlessly incorporated and unquestioned over hundreds or thousands of years. For people working continuously inside the same culture, many of these issues or questions may never arise. Because of different social constructs, individual authenticity is usually less easily perceived, understood, or shared, explicitly and implicitly outside a community, culture, or society than within. To achieve better understanding of one another across cultures, we need to step out of our culture and see from the other's perspective. Once we start working cross-

New Directions for Adult and Continuing Education • DOI: 10.1002/ace

culturally in a global environment, these differences are unmasked. This causes us to question who we are and what we are teaching and learning.

References

Allen, W. C., Hu, H.-L., and Bahr, A. M. (eds.). *Taoism.* New York: Chelsea House, 2005.

Bellah, R. N. *Habits of the Heart: Individualism and Commitment in American Life.* Berkeley: University of California Press, 1985.

Bohm, D. *On Dialogue.* London: Routledge, 1996.

Cranton, P., and Carusetta, E. "Perspectives on Authenticity in Teaching." *Adult Education Quarterly,* 2004, *55*(1), 5–22.

Crouch, S. *The Artificial White Man: Essays on Authenticity.* New York: Basic Civitas Books, 2001.

Dalton, M. M. *The Hollywood Curriculum: Teachers and Teaching in the Movies.* New York: Peter Lang, 1999.

Foucault, M. *Power/Knowledge: Selected Interviews and Other Writings, 1972–1977.* New York: Pantheon, 1980.

Greene, M. *The Dialectic of Freedom.* New York: Teachers College Press, 1998.

Gu, J., and Zhu, Z. "Knowing Wuli, Sensing Shili, Caring for Renli: Methodology of the WSR Approach." *Systemic Practice and Action Research,* 2000, *13*(1), 11–20.

Hansen, D. T. *Exploring the Moral Heart of Teaching: Toward a Teacher's Creed.* New York: Teachers College Press, 2001.

Heckle, P. L. *The Statue of Glaucus: Rousseau's Modern Quest for Authenticity.* New York: Peter Lang, 1991.

Kaufmann, W. A. *Existentialism from Dostoevsky to Sartre.* New York: Meridian Books, 1972.

Lifton, R. J. *The Protean Self: Human Resilience in an Age of Fragmentation.* New York: Basic Books, 1993.

Mezirow, J. *Transformative Dimensions of Adult Learning.* San Francisco: Jossey-Bass, 1991.

Nietzsche, F. W. *Untimely Meditations* (R. J. Hollingdale, trans.). Cambridge: Cambridge University Press, 1997. (Originally published in 1873.)

Tzu, L. *Tao Teh Ching* (J.C.H. Wu, trans.). Boston: Shambhala, 2003.

Lin Lin is a doctoral candidate at Teachers College, Columbia University.

New Directions for Adult and Continuing Education • DOI: 10.1002/ace

7

According to exemplars in adult education, those inner qualities that facilitate the presence of themselves, their learners, and the subject content are the aspects of the genuine self that teachers need to express in the classroom to help bring about learning. This chapter examines what it means to be present, and how teachers develop or come to have this sensibility.

Teaching with Presence

Lloyd Kornelsen

In 2000, after fourteen years of teaching, I took a year off, enrolled in a master's of adult education program, and used it to reflect on my vocation: teaching. The questions and issues I grappled with that year—and the ones that would eventually form the basis for my thesis research—emerged from and were shaped by my years in the classroom. Many of the questions coalesced around, and were epitomized by, a specific phenomenon that grounded the nature and background of my inquiry. I was interested in better understanding what for me was an ultimate classroom teaching-learning event, but one that had eluded finite definition or control. It was the moment when a class or learning group seemed to take on a life of its own, and where participants openly and actively created meaning for themselves, often independent of me, the teacher. During these times I would feel in *flow*, which Csikszentmihalyi and Csikszentmihalyi (1988) describe as forgetting about time, sensing greater inner clarity, losing awareness of self, and yet feeling very present. What caused these teaching-learning episodes to happen was somewhat of a mystery to me. Certainly the appropriate pedagogic methods and techniques were usually present prior to and during these moments, and the learners themselves had to be open and willing to participate. But I also felt that an authentic presence of the teacher, in relationship with his or her students, was critical. So this is what I set out to investigate: What is it about the presence of the person, the teacher, that contributes to the teaching-learning environment?

Through the course of my inquiry, I struggled with what to name this phenomenon, this way of being that facilitated learning. It was not until I actually completed my research that I settled on the term "presence." In

DISCOVER SOMETHING GREAT

NEW DIRECTIONS FOR ADULT AND CONTINUING EDUCATION, no. 111, Fall 2006 © 2006 Wiley Periodicals, Inc.
Published online in Wiley InterScience (www.interscience.wiley.com) • DOI: 10.1002/ace.229

many ways, the phrase "being authentic" could have fit what I was attempting to name. As Cranton (2001) says, "authenticity is the expression of one's genuine Self in the community and society" (p. vii). Expression of the genuine self (or selves) in the community of the classroom was certainly critical and basic to my understanding of what was facilitating "flow-like" teaching moments. But what I was looking to understand was a little more specific: What was it about the genuine self that the teacher needed to express in the classroom to help facilitate learning? What are the particular qualities of the self that need to be expressed and need to be "present?" Put another way, how does the self need to be "there"?

The place I chose to look for an answer to my questions was in conversation with teachers themselves. Using a phenomenological approach, I interviewed three adult educators over a period of several months and observed them in the classroom. These three teachers, whom I refer to as Chris, Jon, and Jill (pseudonyms), are widely recognized as exemplars in classroom teaching practice. All three teach "soft skills" courses and have had diverse teaching careers and experiences. Concomitantly, I reflected on my own teaching practice. Examining my own thinking on the subject and my experience with it, while I was talking to others about it, bound up these two features (my voice and the participant voices) in such a way that they became inextricably linked. So I felt it was important to formally add my voice to the findings. Like the other participants, I have been a teacher of youths and adults and taught diverse courses in a variety of places, from high school to university.

In this chapter, I summarize the significant findings from my study. These findings have to do with what it means to be present, and with how teachers develop or come to have this sensibility. I intersperse in the discussion relevant perspectives from the adult education literature.

Being Present

So what qualities of the teacher's self need to be there, present in the classroom, to help facilitate learning? According to the findings, teachers need to manifest those traits that invite presence: presence of themselves, of their learners, and of the subject-content. Teaching with presence means teaching in a way that encourages openness, imbues vitality, and sometimes abandons order.

Being Open: Helping Learners Be Present. In writing about creating the trusting classroom, Applebaum (1995) says that "a teacher must . . . to a certain degree make him/herself vulnerable to [students] primarily by showing them that he or she is a human being just as they are . . . having faults, weakness, desires, and ambitions" (p. 448). She continues by saying that if a teacher does this, he or she may be able to mitigate power differentials in the classroom and engender a caring, respecting, and trusting environment. This is an environment where learners are freer to risk, challenge, and reciprocate with openness themselves.

New Directions for Adult and Continuing Education • DOI: 10.1002/ace

This quality of openness of self to the group was much emphasized by the research participants. It was variously named: being human, being who you are, being vulnerable and open, being yourself, and being honest.

Jon told a story that captured much of the essence of what was said by all three about what openness means to them and why it is important. At the beginning of a one-day seminar that he was facilitating, he was publicly confronted by one of the participants for what the person believed was a faulty premise in the course content. Jon responded by openly admitting his mistake to the group. Here is his description of what happened next:

> I felt sort of trapped, and there wasn't anything that I wasn't feeling that I wasn't conveying when I said, "You got me." I think they were as aware of the emotion I was feeling as I was, almost. . . . They laughed . . . and the warmth and the comfort in that classroom went up exponentially. . . .
>
> And when I did that [admitting my failure], I think I captured it [their trust]. I think they realized that this guy is honest. And so once I had established that, I think that's possibly why it went so well . . . [because they] saw a human being. Just a human being. A human being who was prepared to be . . . to express what/how he was feeling. I was as vulnerable as I can be. What am I going to do with you? You ruined my thing. I was going to go in this direction. I was going to manipulate you guys to come up with a certain answer, and now what can I do?

According to Jon, what followed was one of his most successful days in teaching. This episode also demonstrates what all three teachers stressed time and again: it is easier for students to trust a teacher and engage with the subject if the teacher is seen to be meaningfully engaging with the subject as well, and open to exploring it together with the students.

Another type of openness that was identified by all participants was openness to students, their lives, their experiences. Participants claimed that such openness helped students feel included and respected; it helped them realize what they already know. In talking about the importance of conscientiously seeking to be open to the life and experiences of her class, Chris told a story of one of her most memorable weeks in teaching. Several years back, the university for which she teaches asked her to teach a five-day communication course in a small northern Manitoba town. All her students were Cree First Nations women. Chris is not aboriginal. She described the first three days as difficult and arduous. Despite her best efforts at being open—inviting participation, asking questions, and seeking to understand her students' perspective—very little seemed to be coming back. No connections between herself, her course material, and her students were being made, or so Chris felt. Midway through the fourth day, the group started participating. Here is how she described what happened on the fifth day, the final day with the group.

New Directions for Adult and Continuing Education • DOI: 10.1002/ace

And at the end of the time, at the end of the five days, one of the ladies came up to me and presented me with a Cree First Nations pin and said the words, "You understand." And that was everything. That was everything to me when they said that. . . . Maybe it was all of that, that I'm not bigger than them; I'm not smarter than them; I didn't come here to tell them about life because I'm different from them, and maybe feared that, that, you know, that I would tell them how it is, when in fact that's not how they live. I didn't do any of that. I asked about them and how we could fit them into the context of my material. And I think they tested me on that by maybe not being too responsive. And by the end they maybe saw that I'm not that.

Ten years later, individual members of this class still visit Chris when they come to the city. Telling the story, Chris stressed the importance of being conscientious—continuing to be curious even if it does not feel as though it is making any difference.

This sounds much like Van Manen (2000) when he talks about teachers needing to be "worryingly mindful" (p. 8)—being singularly open to the call and vulnerability of the other. Or Jarvis, talking about the decision to give up some of one's freedom for the sake of the other, suggesting that openness to the other's humanness will help the teacher generate an environment where students can realize their best educational interest, that of "human becoming" (Jarvis, 1995, p. 35).

Being Vital or "Walking the Talk:" Helping the Subject-Content Be Present. The quality emphasized more than any other was enthusiasm for the subject and having care for and a vitality about the subject that is derived only from experience. It was variously described: having integrity, "walking the talk," being responsible, being enthusiastic. As Jill said, "People . . . know if you're living what you teach, or if you're just teaching what you know, or what you've read. It doesn't matter how well a person delivers it, or how well they speak, it's just not happening. . . . [So] because I care about things that I'm facilitating or instructing, because I try them out, and I experiment with them, and add my own stories, as well, my gifts and my flaws, and stories around my humanness to them, it makes a difference to them [students] because it becomes real."

The suggestion was that if the teacher cares about the subject and if it challenges her or him, then she or he is more able and willing to connect with learners. Moreover, it means that the ego—conscious concern for personal success and acceptance—becomes less important, less apt to interfere with good teaching. Thinking about several of my own positive teaching experiences, I noted: "I challenged them [students] from my heart . . . not by force against them, but from the subject . . . the subject was more important than my ego. This freed me to challenge their assumptions. . . . As I became more engaged with the subject, less concerned about whether I'd be successful, or liked, or how I'd feel at the end of it, or whether the students would interact—the more real and lively the interaction and the class."

New Directions for Adult and Continuing Education • DOI: 10.1002/ace

I saw something similar in the classrooms of the three teachers who participated in my study. For example, my most enduring memory of Jon's class was his integrity. What seemed to matter most to him were the underlying principles of his course. Everything in how he conducted his class seemed to be subject to those principles. Also memorable is my observation of Jill being forthright, early on in class, with her clear and open conviction about her course content. Yet (maybe because of this) both classrooms were filled with interactive energy.

This suggests that respecting the subject, having a personal and intimate connection with it—having vitality about it—is a necessary part of the puzzle of being open to the learners. Both Jill and Chris maintained that when they teach with knowledge that is grounded in experience, it facilitates openness between themselves and their students. This was Jill's assertion: "I'm so passionate [about]. . . the subject matters that I teach, and maybe [that is] why people connect to it." She concluded by saying that the longer she teaches, the more she realizes the subject emerges in the dynamic of the class itself.

This brings to mind Palmer's metaphor (1998) of the subject's presence in the classroom: he depicts the subject as something that rests or sits in the "center" of the classroom around which teachers and students gather and together approach the "great thing" in their midst. Similarly, Vogel (2000) and Bickford and Van Vleck (1997) see the subject having a voice of its own; great teachers recognize it and become one with it. They maintain that to teach well, to teach enthusiastically, means not to take an objective view of the subject, which has the effect of distancing teachers from subject and learners, but to take a subjective approach, which connects all three. Vercoe (1998) adds that to help facilitate a learner's construction of knowledge, it is necessary for the teacher to remake the subject in dialogue every time, with every class.

Living with Chaos. According to the participants, if a teacher in a classroom is other-focused (on students or the subject) and willing to let go of a prescribed agenda in the service of connecting students and subject, the teacher and the group may experience heightened feelings of consciousness and synergy and a sense of physical and emotional well-being. The participants described this phenomenon as a higher state of meaning making and learning. Each told stories of how—in a conscientious quest to be open to the moment, to their students, or to an unfolding revelation in class—they found themselves in this state of "connection." The data showed that if teachers deliberately seek to achieve this state, it does not necessarily happen. It appears to be a by-product of reaching out, of unselfconsciously looking to understand the students, involving them with the class, as one engages with the subject-content. It is not easy; all three participants agreed there is no easy technique or template that guarantees repeatable results. Here is Chris attempting to understand these episodes:

It's not like you've abandoned the curriculum . . . but it's seamless . . . absolutely seamless . . . you're right there with them. You're no longer the instructor. You're just a subtle facilitator, asking questions. *You're not even a guide anymore. You're just there in the conversation, and it moves.* It just keeps moving. You're absolutely aware of everything that's going on around you; and somebody who is looking like they might want to say something and you bring them in and somebody who looks like they're disagreeing and you go there . . . and it becomes a ballet. . . . I don't think I could replicate it because I'm not sure what happened.

This description sounds like Csikszentmihalyi and Csikszentmihalyi's state of flow (1988): forgetting yourself or becoming egoless and focusing on something outside of yourself, bringing more energy to those around. Their depiction of flow is similar to what participants described: a heightened sense of clarity and connectedness to one's surroundings and others. This also seems much like what Heron (1999) describes when talking about the importance of being attuned to the inner experience of people. According to him, being in this state helps teachers respond with more awareness.

In terms of the learning and meaning making that participants reported, this phenomenon corresponds with several similar concepts described in the literature. For example, Bohm and Nichol (1996) picture dialogue as a stream of meaning flowing among and through the group, and Schön (1987) talks of congruence of meaning amid classroom uncertainty. Harris (1987) marvels about the meaning that is revealed in unpredicted and unpredictable circumstances, and Bickford and Van Vleck (1997) describe the music (a metaphor for learning) that happens from something we cannot direct. Chaos theorists such as Wheatley (1999), Capra (1982), and Mossberg (1996) base their thinking on similar conceptualizations. They speak of a chaos–order duality, arguing (not unlike the educators already noted) that if a teacher or group leader lets go of the need to control the chaos (presumably the "messiness" in the classroom), order in the form of higher meaning can emerge.

Both project participants and writers alike talk about the value of reaching out, letting go (of an agenda), and being open in the service of student learning. But what about the quandary, the balancing act, between freedom and order? Where is the balance among being prepared with a thoughtful and organized lesson plan, executing it with knowledge and insight, and remaining open to the spontaneity that emerges from the life of the group and the infinite nature of the subject? All three teachers appeared to have an instinctive sense of balance between the two, knowing when to do what . . . Where does this sensibility come from: experience, intuition, training, or something else? On the basis of how these episodes were described, order (or learning, or convergence of meaning) appears to emerge of its own accord. As Jill and Chris claim, "it just happens," in response to being open and willing to engage with others. This implies that a knowledge of when to do what—guide or let be—takes over in the moment.

New Directions for Adult and Continuing Education • DOI: 10.1002/ace

Jon says he believes that his teaching experience has taught him to more effectively prepare for, recognize, encourage, and guide this process— suggesting that experience plays an important role in developing or exercising this sensibility. As he says, "I have learned to realize that I am there to do nothing but dis-inhibit." Also, he stressed the importance of needing to guide or challenge learners from a base of lived knowledge of the subject-content, implying that our interaction with learners is shaped by our relationship to the subject that we are teaching. The more we know of the subject, the more effective we are in our communication with learners. So whether it is developed through experience, comes from a simple decision to let go, or emerges in a collective quest to understand the subject-content, there appears to be a unique intelligence or skill operating at these times, in these moments.

Becoming Present

From the observations of the project participants, one can say the traits of openness and vitality are most effectively exercised (present) when teachers feel free to be themselves. This echoes the writings of adult educators such as Apps (1996), Cranton and Carusetta (2004), Heron (1999), and Palmer (1998). However, it may take years for teachers to develop the confidence to trust themselves as well as the insight to realize the importance of doing so. Moreover, it is an ongoing daily teaching challenge; unlike a learned skill (such as how to ask an open question), which stays with you, exercising an inner quality (being open) is a personal choice and commitment a teacher makes every time she walks into a classroom. So how, then, is this ability or disposition learned or developed?

Moving from Techne to Phronesis. In describing their growth as teachers and how their view of teaching has changed, Jon, Chris, and Jill revealed some remarkably similar sentiments. They all talked about how at the beginning of their teaching lives they were preoccupied with technique, structure, and course content. But over the years the priority shifted to "the people": bringing them (learners) in, bringing themselves (teachers) in, and letting the subject-content "be." They suggest that as they became more confident in themselves and their teaching skills, it freed them to be themselves and shift their attention to the human interaction they claimed was at the heart of the teaching–learning process. It was not that techniques and methods no longer mattered; they did. But the emphasis shifted, from their implementation to their purpose.

This transition from a way of doing to a way of being corresponds to Dunne's description (1993) of the two forms of Aristotelian knowledge, *techne* and *phronesis*. Techne is knowledge possessed by a maker and suggests sovereignty over; phronesis is knowledge that is personal and suggests communal engagement with. Dunne argues that teaching (or any form of human interaction) cannot be reduced to technique, because teaching is not a process of making objects but a practice of engaging in human interaction. He says that this calls for teachers to bring qualities of mind, character, and practice transcending skillful application of technique. To teach effectively, Dunne says, the practice of techne is indispensable. However, to ensure that the techniques are deployed in right relation to the right person in a given situation, what teaching calls for is knowledge of phronesis. Phronesis must underlie techne.

All three teachers experienced a similar transition in their approach to teaching, from teaching with emphasis on techne to teaching with an aptitude that might itself be called phronesis. What this implies is that teachers may develop this knowledge, this phronesis, with time, experience, and confidence. (This is corroborated by recent research by Cranton and Carusetta, 2004.) It also suggests that being self-confident and skilled in teaching techniques (having the knowledge of techne) may presage teaching with presence, teaching with a focus on human interaction.

It should be noted that even though all three teachers spoke of similar transitions in their approach to teaching, how their phronesis was manifested in the classroom differed considerably. Their teaching styles were quite diverse, which is what one would expect; as Cranton (2001) says, if we teach in a way that is true to ourselves, we will teach differently from each other. There is no behavioral template for being present.

Being Careful and Committed. Another important part in acquiring this disposition of presence was revealed not in what I found but in what I did *not* find. It suggests an unspoken, even unconscious assumption on my part about the study. In retrospect, I think I hoped to find in the three teachers I interviewed some magic bullet—an unusual inner quality, a profound insight, or a watershed experience—that would singularly and extraordinarily account for their classroom success and their teaching presence. I did not find that magic bullet, at least not the way I thought I would. What I encountered instead were three ordinary people, very much like me, who offered variations, extensions, confirmations, and several disconfirmations of my own thinking on the subject.

At the end of the interview process, I felt mildly disappointed. I had not discovered or uncovered a magic solution to the puzzle of excellence in teaching. What I did find were three people who had extraordinary commitment to their vocation and their students, and it was this that I found exceptional. Here is Jon encapsulating what all three emphasized time and again: "It's not about me and my needs, it's what they [students] need and want . . . to help my students live lives that are more suited to who they are." Chris, Jill, and Jon were teachers to whom teaching mattered and to whom being open and vital in the classroom was a necessary, conscientious, and ongoing process. There was remarkable congruence between what they said they believed in and who they were in the classroom.

I learned something that perhaps has more significance and value for teachers than a magical solution: that teaching with presence is hard work. There is no easy way, gimmicky or otherwise. Talking with the three teachers and watching them in action confirmed the proposition I have struggled with throughout my teaching career: teaching with presence takes effort and commitment day in and day out. It is not some skill or quality one can possess once and for all. Whether to be present or not is a decision to be made every time one walks into a classroom. As Van Manen (1990) claims in talking about pedagogy, in every situation a teacher must continuously redeem, retrieve, regain, and recapture. Walck (1997) describes the vocation of the teacher as an unending process of building a frame of mind, one of freedom and openness to all that can and will happen.

Maybe the inner quality that is at the root of good teaching, the one underpinning everything else, is commitment. For a teacher to exercise qualities that help in being present, there must first be a commitment, ongoing care for the well-being of one's students and the value of learning. This affirms what Paulo Freire said thirty years ago, and what many adult educators have suggested since then: "To be a good educator you need above all to have faith in human beings. You need love" (cited in Vercoe, 1998, p. 56).

Conclusion

Teaching with presence entails teaching in a way that facilitates the presence not only of the teacher but of the learner and the subject-content as well. It may mean having to accept—and perhaps even embrace—some chaos in the classroom. For teachers to exercise these qualities and do so effectively, they must be committed to teaching and have a sense of self-confidence and freedom. The confidence to be present, and to make it a teaching priority, develops with experience.

To help facilitate a connective, flowlike learning environment, a teacher needs to be open to meaningful student participation, and not fear the messiness or chaos that results. Teachers need to be themselves and trust themselves. They can be more effective if they worry less about fitting into some prescribed teaching template or style and concern themselves more with the well-being of their students and the integrity of the subject-content. By learning more, and learning alongside their students in the classroom, the learning environment will become vitalized. Their relationship to the subject affects how it is revealed in the classroom.

Teaching with presence cannot be easily judged, quantified, or measured. There is knowledge of teaching—an intelligence that is subjective and interpersonal in nature and often acquired through experience—that cannot be evaluated in the same way as the technical aspects of teaching. This calls for a means of investigation falling into the realm of what Van Manen (1990) calls human science. It is an approach to research that seeks to understand the world through the lived experiences of human beings. Accordingly, to more accurately determine the effect of a teacher's presence in the classroom and its meaning for learning, one has to take into account the descriptions of those people who live that experience. An attempt to objectify this may discourage a way of being that can happen only in an environment of openness and freedom.

References

Applebaum, B. "Creating a Trusting Atmosphere in the Classroom." *Educational Theory*, 1995, 45(4), 443–452.
Apps, J. W. *Teaching from the Heart.* Malabar, Fla.: Krieger, 1996.
Bickford, D. J., and Van Vleck, J. "Reflections on Artful Teaching." *Journal of Management Education*, 1997, 21(4), 448–472.
Bohm, D., and Nichol, L. (eds.). *On Dialogue.* London: Routledge, 1996.
Capra, F. *The Turning Point: Science, Society, and the Rising Culture.* Toronto: Bantam Books, 1982.
Cranton, P. *Becoming an Authentic Teacher in Higher Education.* Malabar, Fla.: Krieger, 2001.
Cranton, P., and Carusetta, E. "Developing Authenticity in Teaching." *Journal of Transformative Education*, 2004, 2(4), 276–293.
Csikszentmihalyi, M., and Csikszentmihalyi, I. S. (eds.). *Optimal Experience: Psychological Studies of Flow in Consciousness.* New York: Cambridge University Press, 1988.
Dunne, J. *Back to the Rough Ground: 'Phronesis' and 'Techne' in Modern Philosophy and in Aristotle.* London: University of Notre Dame Press, 1993.
Harris, M. *Teaching and the Religious Imagination.* San Francisco: HarperSanFrancisco, 1987.
Heron, J. *The Complete Facilitators Handbook.* London: Kogan Page, 1999.
Jarvis, P. "Teachers and Learners in Adult Education: Transaction or Moral Interaction?" *Studies in the Education of Adults*, 1995, 27(1), 24–35.
Mossberg, B. "Teaching for Turbulence." *National Teaching and Learning Forum*, 1996, 5(3), 5–7.
Palmer, P. J. *The Courage to Teach: Exploring the Inner Landscape of a Teacher's Life.* San Francisco: Jossey-Bass, 1998.
Schön, D. A. *Educating the Reflective Teacher.* San Francisco: Jossey-Bass, 1987.
Van Manen, M. *Researching Lived Experience: Human Science for an Action Sensitive Pedagogy.* London, Ont.: Althouse Press, 1990.
Van Manen, M. "Moral Language and Pedagogical Experience." *Journal of Curriculum Studies*, 2000, 32(2), 315–327.
Vercoe, A. "The Student–Teacher Relationship in Freire's Pedagogy: The Art of Giving and Receiving." *New Zealand Journal of Adult Learning*, 1998, 26(1), 56–73.
Vogel, L. "Reckoning with Spiritual Lives of Adult Educators." In L. M. English and M. A. Gillen (eds.), *Addressing the Spiritual Dimensions of Adult Learning: What Educators Can Do.* San Francisco: Jossey-Bass, 2000.
Walck, C. L. "A Teaching Life." *Journal of Management Education*, 1997, 21(4), 473–482.
Wheatley, M. J. *Leadership and the New Science: Learning About Organization from an Orderly Universe.* San Francisco: Berrett-Koehler, 1999.

LLOYD KORNELSEN *is a private consultant and facilitator, and a full-time instructor at the University of Winnipeg in Manitoba, Canada.*

8

This chapter summarizes and integrates the main themes in the volume: examination of authenticity through a critical lens of power, culture, and gender; and consideration of authenticity as an essentially humanist endeavor.

Integrating Perspectives on Authenticity

Patricia Cranton

I think about authenticity a lot. Every year, I teach a course called Becoming an Authentic Teacher and another called Imagination, Authenticity, and Individuation in Transformative Learning. I continue to do research on authenticity in teaching; currently it takes the form of narrative inquiry as I work with teachers in higher education and adult education over a three-year period. They tell me stories about their practice, and together we wonder about what authenticity means to them. Recently, I interviewed an educator from the Canadian military who said he did not feel constrained in his practice. If he did not agree with something, there were procedures by which he could change the thing. Then, in a response to another comment of mine, he said, "I have no values when I teach; everything is predetermined and tested, and I simply teach it." Yet he feels authentic.

I also worry about authenticity a lot. I worry on several levels. I am not quite sure what authenticity is. As Russell Hunt says in Chapter Five, it is easier to think about what it isn't than what it is. Inevitably, my students, following an "expression of the genuine self in the community" definition (Cranton, 2001), get around to saying you can be an authentic teacher who is not a good teacher (Hunt also touches on this); but worse, they push this further and come up with the idea of authentically evil people who are genuine in their belief that it is their duty to cleanse the world of some group or political system. On another level (perhaps the level of premise reflection), I wonder why I am interested in this topic in the first place. Yesterday, I had this chapter in my mind as I was stacking firewood. On the news

NEW DIRECTIONS FOR ADULT AND CONTINUING EDUCATION, no. 111, Fall 2006 © 2006 Wiley Periodicals, Inc.
Published online in Wiley InterScience (www.interscience.wiley.com) • DOI: 10.1002/ace.230

was Israeli Prime Minister Ariel Sharon's massive stroke, 120 Iraqis killed in suicide bombings earlier in the day, and twelve miners in West Virginia dead. Here I am, thinking about how individual teachers can or should be authentic in their practice. What does it matter if Francis feels depressed after his class (John M. Dirkx's Chapter Three) or Katherine A. Frego's students feel anxious about a test (Chapter Four)? Shouldn't I (and all of us in adult and higher education) be thinking about more important topics? Shouldn't we be looking for ways to make the world a better place, or at least teaching our students how to make the world a better place?

But perhaps this does take us back to authenticity in teaching after all. As Dirkx emphasizes in Chapter Three, authenticity is founded on continuing deep development of a sense of self. He draws a parallel between the journey toward authenticity and individuation, a Jungian concept that describes how we differentiate ourselves from the collective of humanity while at the same time finding our place in that collective. Becoming authentic is, in many ways, individuation. It is not only being genuine but understanding what genuine means in a deep way for ourselves, and this involves critically questioning the world outside of ourselves. To differentiate ourselves from the collective psyche, we need to come to terms with which social norms have been uncritically assimilated. What do we really believe? What do we hold dear? What resonates with us, and what does not? It is in this place, I think, where we can see that being (or becoming) authentic has the potential to lead us to rise up against the evils of the world. We must, in being true to ourselves, object to and work against what we see as wrong.

In Chapter One, Brookfield critiques authenticity through the lens of power. Following the work of Ian Baptiste, he argues that educators need to conceal their intentions from learners in order to teach critically—to raise uncomfortable and challenging viewpoints. The students would leave, he suggests, if they were to know our true goals. In this critique, Brookfield presents an apparent paradox between being authentic and exercising power in order to challenge learners to think critically about important issues in the world. But if we all, educators and learners, are on the journey of authenticity and individuation, whether it is conscious or not (as Jung proposes), then we are all engaged in seeing how we are different from and simultaneously the same as the collective of humanity. The educator is helping the learner become more authentic (which is part of Jarvis's definition of authenticity; 1992). No one, including the educator, knows exactly where this will lead, and there really are no hidden agendas. We are in the business of exploring and learning about the world, and through that about ourselves.

This theme is apparent elsewhere in the volume. Leona M. English, in Chapter Two, suggests that our attempts to pay attention to women's learning serves to further marginalize women by describing them as relational, collaborative, and "nice." In other words, in our attempt to help others develop, we can bypass their authenticity, and our authenticity as educators, by focusing on only one dimension of their being and responding to it. Women's issues become divorced from the social and cultural arena in which they arise, says English. This parallels, in my mind, Brookfield's argument that by being responsive to learners' expressed needs and wants we avoid challenging them to examine their own assumptions, and thereby not help them see themselves as separate from the collective—a vital aspect of becoming authentic.

Similarly, Lin, writing in Chapter Six about cultural differences in people's responses to authenticity, worries about stereotyping. If we say that Asian learners have a certain reaction to authenticity in teaching or that Asian and North American cultures do not have the same expectations and values regarding their teachers, we have the problem of either being essentialist in grouping together all individuals from one culture and saying they are alike in some fundamental way or ignoring broader sociocultural perspectives. Examining authenticity through the lens of critical theory, power, feminism, or sociocultural issues leaves us with this struggle every time. It seems we need to help learners articulate and question their own assumptions about power, gender, culture, and learning, and help them find their own way to be and feel empowered in a broad social and global context that includes all of these complexities. Educators need to be coexplorers; by not doing so, we too are trapped in the lens through which we see the world, including our own assumptions about power, gender, culture, and learning.

Brookfield struggles with what he sees as a contradiction between being both authentic and true to his agenda as a critical teacher; Hunt, in Chapter Five, examines a similar problem but differently. He is interested in how institutional constraints inhibit authenticity in teaching. Among those institutional constraints are, inevitably, social expectations and assumptions about what an educator should be doing. Hunt chooses an alternative way to work with (or against) those expectations. He too is a critical teacher; his goal is to help students challenge and critically examine pervasive points of view. Yet when his students follow a task to a place Hunt and his colleagues do not intend (and a place he feels is simply wrong), he realizes that to steer the students' investigation would be to violate his own convictions about the best way for students to learn. Both Brookfield and Hunt come to the same conclusion—that such a situation is not easily resolvable, or perhaps not resolvable at all—and it is helpful to the reader of this volume to see two quite distinct ways of responding to the irresolvable.

It is interesting that more than half of the chapters in this volume address, in some way, power issues, including contradiction between what the educator thinks is best and what the educator thinks the learner thinks is best. The second major theme in the volume centers on humanism, relationships, and development of the self. At first glance at least, the study of authenticity in teaching appears to be a humanist endeavor. It is about self-awareness, self-development, and genuine relations and communication between self and other. When Hunt struggles with the discrepancy between what he intends and what students choose to do, he goes with what students choose, so in this way he steps away from his agenda and fosters student independence. He bridges the two themes in the volume by putting student choice ahead of his own, even though his goal is to develop students' critical thinking.

This is an ongoing tension in working toward authenticity in teaching. What do we do when our expectations, assumptions, or values differ from those of our learners? The critical theorist challenges the learner, perhaps even to the point of using manipulation or coercion. The humanist focuses on self—teacher-self, student-self, and the relationship between the two. Most of us try to work in some place in between, where we challenge, respect, and care for our students. Frego, in Chapter Four, illustrates this well. She says quite clearly that no relationship can be positive and productive if the participants are not genuine, or if the intent is to manipulate or deceive. She strives to develop relationships based on reciprocal honesty and good intention, relationships that focus on learning through appropriate caring. Frego acknowledges the educator's power but chooses to create trust. Although this may make her feel vulnerable, openness takes priority over any risk she might face. She is the first to communicate "I care about you; you are important" without requiring or even expecting that students will reciprocate.

Lloyd Kornelsen, in Chapter Seven, echoes Frego's emphasis on trust and openness to the point of making oneself vulnerable. The participants in his study talked about being human, being who you are, being vulnerable and open, and being honest. Kornelson frames his work with the idea of "being present." The teacher, the learners, and the subject area or content of the learning are strongly present; teachers are other-focused (on the students and the content) and are willing to let go of their own expectations and agendas even if chaos in the classroom results. Again, the contrast with a pedagogy of coercion is striking.

Frego and Kornelsen write about the educator's vulnerability in the process of being authentic, but it is John M. Dirkx, in Chapter Three, who takes an in-depth look at the personal quest for authenticity. Teaching, for Dirkx, is intimately bound up with who we are as persons, and the development of authenticity depends on our willingness to "muck around in the dark, messy, unpredictable world of the unconscious." It is a symbolic quest, a demanding trek, one that needs to be approached with respect, wonder, humility, and love.

Becoming an authentic teacher, as Dirkx suggests and as all the authors in this volume reflect, requires a profound commitment of time and intellectual and emotional energy. I consciously question my practice all the time, through my teaching journals, discussion with colleagues, discussion with students, and writing and research. Reading Dirkx's chapter also allows me to realize that there are unconscious facets of the process—facets that show up in how I am drawn to stories and movies, in my dreams, and in my sometimes seemingly inexplicable emotional reactions to events in my teaching.

There are times in my practice that stay in mind with vivid detail as I continue to work through what they mean in relation to my authenticity—the student who demanded that I "teach" her (tell her what to do), given that she had paid good money for the course; the group who chose not to learn about research methodologies in a research methods course; the student who felt free to talk and cry about her father's death for the first time (several years after his death) in my class. It is my hope that the chapters in this volume will engage readers in similar reflection about their own teaching, and perhaps encourage them to see their work in some meaningful new way.

References

Cranton, P. *Becoming an Authentic Teacher in Higher Education.* Malabar, Fla.: Krieger, 2001.

Jarvis, P. *Paradoxes of Learning: On Becoming an Individual in Society.* San Francisco: Jossey-Bass, 1992.

PATRICIA CRANTON *is a visiting professor of adult education at Pennsylvania State University at Harrisburg.*

INDEX

Adult education, women's learning and, 18–20

Articles: "Authenticity and Imagination," 27–39; "Authenticity and Power," 5–16; "Authenticity and Relationships with Students," 41–50; "Cultural Dimensions of Authenticity in Teaching," 63–72; "Institutional Constraints on Authenticity in Teaching," 51–62; "Integrating Perspectives on Authenticity," 83–87; "Teaching with Presence," 73–82; "Women, Knowing, and Authenticity: Living with Contradictions," 17–25

Authentic teaching: awareness of others and, 1–2, 65–68; caring and, 42–44, 46; commitment and, 38, 81, 86; contextual influences and, 68–70, 85; credit system and, 56; critical reflection and, 30–31, 33, 70–71; cross-cultural issues and, 65–67, 85; cultural issues and, 68, 71–72, 85; curricular structure and, 55; defining, 1, 52–53, 83; dilemmas regarding, 62; "good teaching" policies and, 56–57; identity and, 53; imagination and, 32; institutional contexts and, 53–54; institutional rules and, 55; integrity and, 44, 51, 53; learning and, 41; multiculturalism and, 69; opposing institutional constraints and, 58–59; physical constraints and, 54–55; policy constraints and, 56–57; respect for privacy and, 44; role expectations and, 57–58, 70; self-awareness and, 1–2, 52, 65–68, 84; self-knowledge and, 29, 33–34; social constraints and, 57–59, 85; soul work approach to, 34–38; specific experiences and, 34–35; structural constraints and, 54–56; symbolic approach and, 33, 37; teaching from the heart and, 28–29, 37; tenacious humility and, 68; tension in working toward, 86; unconscious and, 37, 86. See also Authenticity, of teacher; Presence in teaching

Authentic teaching, soul work and: journaling and, 34; self-awareness and, 32–33; self-knowledge and, 29, 34–38; specific experiences and, 35; symbolic approach and, 35–36; use of fantasy and, 36–37

Authenticity: awareness of context and, 1–2, 68–70; awareness of others and, 1–2, 65–68; being and appearance and, 70–71; critical self-reflection and, 1–3, 52; cross-cultural issues and, 64–67, 85; defining, 1, 52–53, 83; existential philosophers and, 69; five dimensions of, 1; genuineness/helpfulness and, 18; individuation and, 66–67, 84; judging, 66; as modernistic sentiment, 18; openness and, 24; personal growth and, 71; pure self and, 18; relationships and, 1–2; self-awareness and, 1–2, 52, 65–68, 84; self-knowledge/understanding and, 29, 34; social nature of, 52; women's learning and, 18. See also Authentic teaching; Authenticity, of teacher

Authenticity, of teacher: being helpful to students and, 6; communication and, 71; congruence of words/actions and, 7–8; critical indicators of, 6–11; dimensions of, 1–2; full disclosure of criteria and, 89; indicators of, 6–11; personhood of teacher and, 10–11; power differentials and, 43, 74, 84; responsiveness to students and, 9–10; self-awareness and, 1–2, 52, 65–68, 84; soul and, 31; teacher as ally/authority and, 5–6; teacher credibility and, 5–6; teacher effectiveness and, 5; teaching for intellectual development and, 16; women's learning and, 22–24, 84–85. See also Authentic teaching; Effectiveness, of teacher; Power, and teacher authenticity; Presence, in teaching

"Authenticity and Imagination," 27–39

"Authenticity and Power," 5–16

"Authenticity and Relationships with Students," 41–50

Awareness of context, as dimension of authenticity, 1–2

Awareness of others: authentic teaching

89

tions and, 44–45; student needs and, 44; teacher authenticity and, 9–10, 71; teacher openness and, 74–76; teaching quality and, 28; trust and, 5–6, 43; trust building in, 5–6, 74; valuing individuals and, 45–46

Teachers: as allies/authorities, 5; caring about subject by, 76–77; as counterfeit critical thinkers, 7; emotional experiences of, 28–29, 32–33; phony responsiveness of, 7; presence of, 74, 77–79; as spuriously democratic, 7; traits for presence in, 74–79; trust building by, 5–6, 74; walking the walk by, 76–77. See also Authenticity, of teacher; Effectiveness, of teacher; Trustworthiness, of teacher

Teaching: emotional dimension of, 28; flow and, 73, 78, 81; human science and, 81; soul and, 29–30; teacher-student relationship and, 28; women's learning and, 22–24. See also Authentic teaching; Presence, in teaching

"Teaching with Presence," 73–82

Techne-phronesis shift, 79–80

Trust: teacher effectiveness and, 5; teacher openness and, 74–76, 86; teacher-student relationship and, 43. See also Trustworthiness, of teacher

Trustworthiness, of teacher: congruence

of words/actions and, 7–8; full disclosure of criteria and, 8–9; personhood of teacher and, 10–11; responsiveness to students and, 9–10; teacher authenticity and, 6. See also Teacher authenticity

Vitality, in teaching, 76–77, 79

Vulnerability. See Openness, in teaching

Walking the walk, by teachers, 76–77

"Women, Knowing, and Authenticity: Living with Contradictions," 17–25

Women's learning/knowing: adult education and, 18–20; circles and, 22; contextualizing, 23; essentialization of, 19; feminist separate view of, 20; feminist studies and, 19–20; feminist third-wave view of, 20; inclusiveness and, 22; openness and, 24; origins of notion of, 18–19; postfoundationalist studies and, 20–21; power issues and, 21–24; self-revelation and, 22; simplified portrayal of, 17; social constructionism and, 19; teaching/learning implications and, 22–24, 84–85; verbal participation and, 22–23; Women's Ways of Knowing and, 18–19; writings on, 17